Topology

Stefan Waldmann

Topology

An Introduction

 Springer

Stefan Waldmann
Julius Maximilian University of Würzburg
Würzburg
Germany

ISBN 978-3-319-09679-7   ISBN 978-3-319-09680-3   (eBook)
DOI 10.1007/978-3-319-09680-3

Library of Congress Control Number: 2014945348

Mathematical Classification Code: 54-XX, 54-01

Springer Cham Heidelberg New York Dordrecht London

Printed on acid-free paper

Springer is part of Springer Science+Business Media (www.springer.com)

To Robert, Sonja, Richard, Silvia, and Viola

Preface

These lecture notes grew out of notes I prepared for my lecture *Topology* held in Erlangen in Summer Term 2012. The course is aimed at bachelor students in their second year being familiar with the basic notions from calculus. The purpose of this small course is to give some first introduction to the notions of general (or point set) topology as they are needed in many other areas of mathematics. Of course, there are many excellent textbooks on topology available. However, the aim of these notes as well as of the lecture itself is to give bachelor students after their first year a *minimal* amount of topology needed to continue with more advanced topics in a mathematics (or physics) programme but still providing detailed proofs. Moreover, the idea is to require only very basic preliminary knowledge as offered by the introductory calculus and linear algebra courses.

The text is self-contained and provides many exercises which will enable the student to work through these notes on her own. Alternatively, the text may serve as a companion for a small lecture on topology.

The first four chapters can be seen as the core of the theory which every mathematics student and hopefully also some physics students should be exposed to. The remaining two chapters give a certain personal preference: I have chosen to put some focus on possible applications in functional analysis. This explains why the last two chapters are on continuous functions as well as on Baire's Theorem in different formulations. On the other hand, I have omitted other important concepts like the fundamental groupoid or topological groups and their continuous actions due to the lack of time and space.

The participants of the original lecture in Erlangen showed great patience with the first versions, not only of these notes but also with the lecture itself. I would like to thank all of them for their comments, remarks and suggestions, which all found their way into these notes in one form or the other. In particular, I would like to mention here Alexander Spies for numerous corrections and a careful proofreading of the entire manuscript. Moreover, I am much obliged to Florian Unger for taking care of the *LATEX*-files and all the typing of the first version of the draft. Without his help, the manuscript would never have been finished. It is a pleasure to thank Karl-Hermann Neeb for various discussions on the pedagogical aspects of teaching of topology. The anonymous referees pointed out many weak points in the original manuscript helping to improve it in many places. Their comments and remarks are much appreciated. Last but not least, I would like to thank my family for the patience and support throughout: without this the book would never have been possible.

Würzburg, June 2014   Stefan Waldmann

Contents

Symbols

$M, N, \ldots$	Sets
2^M	Power set of M
(M, d)	Metric space with metric d
$\mathbb{S}^n$	n-Sphere in $\mathbb{R}^n$
$\mathbb{T}^n = \mathbb{S}^1 \times \cdots \times \mathbb{S}^1$	n-Torus
$\mathrm{B}_r(p)$	Open ball centred at p with radius r
$\mathfrak{U}(\mathfrak{p})$	Set of all neighbourhoods of p
$(p_n)_{n \in \mathbb{N}}$	Sequence of points
$(\mathrm{M}, \mathcal{M})$	Topological space with topology $\mathcal{M} \subseteq 2^M$
$\mathcal{O} \subseteq M$	Open subset of topological space
$A \subseteq M$	Closed subset of topological space
$\mathcal{S}, \mathcal{B} \subseteq \mathcal{M}$	Subbasis or basis of a topology
$\mathcal{M}\|_N$	Subspace topology for $N \subseteq M$
A°	Open interior of a subset $A \subseteq M$
∂A	Boundary of a subset $A \subseteq M$
A^{cl}	Closure of a subset $A \subseteq M$, also $\overline{A}$
$\mathscr{C}(M, N)$	Continuous maps from M to N
$\mathscr{C}(M) = \mathscr{C}(M, \mathbb{C})$	Continuous complex-valued functions on M
$\mathfrak{C}, \mathfrak{D}, \ldots$	Categories
$\mathsf{Obj}(\mathfrak{C})$	Objects of $\mathfrak{C}$
$\mathsf{Morph}(a, b)$	Morphisms from a to b for $a, b \in \mathsf{Obj}(\mathfrak{C})$
top	Category of topological spaces
Top	Category of Hausdorff spaces
$\mathfrak{C}(\mathfrak{p})$	Connected component of $p \in M$
$\Pi(p)$	Path-connected component of $p \in M$
T_i	Separation property i
$\mathscr{C}_b(M)$	Bounded continuous complex-valued functions on M
$\|f\|_\infty$	Supremum norm of a function f
C	The Cantor set
$M = \Pi_{i \in I} M_i$	Cartesian product of sets or topological spaces

$\mathrm{pr}_i : M \longrightarrow M_i$	Projection onto i-th factor
$\sim$	Equivalence relation on M
$M/\sim$	Set of equivalence classes
$\pi : M \longrightarrow M/\sim$	Quotient map
$\Delta_M \subseteq M \times M$	Diagonal in $M \times M$ with diagonal map Δ
$\Phi : G \times M \longrightarrow M$	Group action of G on M
$G \cdot p$	Orbit of $p \in M$
M/G	Set of orbits, orbit space
(U, x)	Topological chart with local coordinates
$\mathbb{RP}^n, \mathbb{CP}^n$	Real and complex projective space
$(I, \preccurlyeq)$	Directed or partially ordered set
$(p_i)_{i \in I}$	Net of points $p_i \in M$
$\lim_{i \in I} p_i$	Limit point(s) of a net
A^{scl}	Sequential closure of a subset $A \subseteq M$
$\mathfrak{F}$	Filter on M
$f_* \mathfrak{F}$	Push-forward filter
$\mathfrak{F}_A$	Trace filter on a subset $A \subseteq M$
$\{\mathcal{O}_i\}_{i \in I}$	Open cover of M
$K \subseteq M$	Compact subset of M
$(M^*, \mathcal{M}^*)$	Alexandroff compactification of $(M, \mathcal{M})$
$\mathcal{A}$	C^*-Algebra or subalgebra of $\mathscr{C}(M)$
$\{K_n\}_{n \in \mathbb{N}}$	Exhausting sequence of compact subsets
$\mathcal{O}(X)$	Holomorphic functions of open subset $X \subseteq \mathbb{C}$
$\|\cdot\|_K$	Supremum norm over compact $K \subseteq M$
p	Seminorm on a vector space
$\mathrm{B}_{\mathrm{p},r}(x)$	Open ball with respect to seminorm p
$\mathscr{B}_{\mathrm{loc}(M)}$	Locally bounded functions on M
$\mathrm{dist}(p, A)$	Distance of point p to closed subset A
$\mathrm{supp}\, f$	Support of a continuous function
$\mathscr{C}_{\mathrm{K}}(M)$	Continuous functions with support in K
$\mathscr{C}_0(M)$	Continuous functions with compact support

Chapter 1
Introduction

The basic idea of topology is to axiomatize the properties of open subsets in $\mathbb{R}^n$: one can take arbitrary unions and finite intersections of open subsets to obtain again open subsets, and the empty set as well as the total space are open, too. This already provides the precise definition of a topology, i.e. a collection of subsets of a set M which should be regarded as "open". A key feature of continuous functions on $\mathbb{R}^n$ is that a function is continuous iff the inverse images of open subsets are again open. This observation can then be turned into the definition of continuity for maps between topological spaces. With these two definitions one has already the core ingredients to develop the theory of general topological space, called general topology or point set topology.

In these notes we will give some first introduction to topology. The field of topology is a very classical part of mathematics which every student should be exposed to at least once. Many of the notions we will encounter will be helpful if not crucial to understand more advanced topics in mathematics, in particular in functional analysis, differential geometry, and algebraic topology, to name just a few. Nevertheless, these notes provide only an overview and it will remain a task for the reader to continue at other more advanced texts whenever this is needed.

A guiding principle throughout this text will be the situation in $\mathbb{R}^n$ which we would like to generalize somehow as far as possible, or at least very far. This way we will obtain topological spaces sharing not too many features with $\mathbb{R}^n$ but providing some new and sometimes very unexpected properties. One major difficulty in general topology will be to handle this zoo of new notions and effects in a reasonable way. It will be very important to understand which of the new features of topological spaces occur in many other contexts and have some general interest and which of them provide only rather pathological behaviour. We will only discuss a certain small part of the possible features and their mutual relations.

One of the major advantages of topology is that after passing to the general framework, many theorems become very easy, sometimes almost trivial. Of course, some other difficulties show up and make the whole theory interesting. A good example for the former is the continuity of the composition of two continuous maps,

© Springer International Publishing Switzerland 2014
S. Waldmann, *Topology*, DOI 10.1007/978-3-319-09680-3_1

the latter is illustrated e.g. by the complications caused by the difference of continuity using nets or filters and sequential continuity.

The material is presented in several chapters, the first four of them can be seen as the core of the course. After introducing topological spaces and continuous functions we discuss many additional features and properties in Chap. 2. Here a particular focus is on the notions of neighbourhoods, interior and closure of subsets, the connectedness properties of topological spaces, and the separation axioms. While both for connectedness and separation there exists a zoo of notions we have focused here on the very basic and most important ones. Then Chap. 3 deals with some fundamental constructions of topological spaces being build out of other topological spaces. Here the topological manifolds point into the direction of differential geometry, while final and initial topologies are used in many places, in particular in functional analysis but also in various quotient constructions in algebraic topology and differential geometry. Chapter 4 contains a detailed analysis of the notions of convergence. Here we need new concepts if the topological spaces do no longer satisfy the first axiom of countability: nets and filters enter as the more generic and more powerful notions compared to sequences. In Chap. 5 we introduce the concepts of compactness and sequential compactness. Compact spaces usually show a much simpler and nicer behaviour in many ways, so understanding them in quite some detail usually provides a good starting point when moving on to the non-compact case for any sort of problem. One of the most important theorems in general topology is Tikhonov's Theorem stating that a Cartesian product of compact spaces will be again compact. The applications of this theorem are indeed overwhelming and we indicate only few of them in the discussion of Tikhonov's Theorem.

After these four chapters, the following two contain some more particular topics: in Chap. 6 we consider continuous functions and show the four main theorems on continuous functions: Urysohn's Lemma and Tietze's Theorem provide the existence of many continuous functions with certain properties once the topological space enjoys the appropriate separation axioms. The Stone-Weierstraß Theorem gives a simple and easy to handle criterion when arbitrary continuous functions can be approximated by simpler ones. Finally, the Arzelà-Ascoli Theorem characterizes the compact subsets of continuous functions. Chapter 7 continues with a detailed discussion of Baire's Theorem. The notion of meager subsets generalizes the nowhere dense subsets in a very useful way. The statement of Baire's Theorem has many applications, mainly in functional analysis. We conclude this short introduction to topology with some rather surprising and amusing facts on the discontinuities of functions which are obtained by limit processes from continuous ones.

In a small appendix we collect some useful formulas from set theory and give a precise formulation of Zorn's Lemma and the Axiom of Choice as it is used in the main text. However, this is far from being an introduction to set theoretic concepts but just a cheat sheet instead.

It is clear that such an important and classical field of mathematics like topology can hardly be covered by such a small book. For a more detailed approach and also for several other topics which are not covered in the following, we would like refer to the textbooks [3, 13, 17, 27, 32]. Here one finds additional references, further details and examples as well as many more advanced topics of topology. After mastering the general aspects of topology as presented here, one may want to continue in various directions: algebraic topology including homotopy and sheaf theory is treated in textbooks like [4, 5, 8, 31, 33], differential topology and topological manifolds are discussed in e.g. [12, 22]. The whole world of functional analysis is clearly to vast to describe, but first introductions can be found e.g. in [20, 28, 29, 35] to name only a few.

As a last word of guidance, it should be clear that whenever one wants to learn some topic in mathematics some labour is needed: there are plenty of exercises which help to deepen the understanding of the core text. They also provide some additional examples and open the horizon for further developments of the theory.

Chapter 2
Topological Spaces and Continuity

Starting from metric spaces as they are familiar from elementary calculus, one observes that many properties of metric spaces like the notions of continuity and convergence do not depend on the detailed information about the metric: instead, only the coarser knowledge of the set of open subsets is needed. This motivates the definition of a topological space as a set together with a collection of subsets which are declared to be the *open subsets*. The precise definition requires crucial properties of open subsets as they are valid for metric spaces. Having managed this axiomatization of "openness" it is fairly easy to transfer the notions of continuity and neighbourhoods to general topological spaces.

Already for metric spaces and now for general topological spaces there are several notions of connectedness which we shall discuss in some detail. New for topological spaces is the need to specify and require separation properties: unlike for a metric spaces we can not necessarily separate different points by open subsets anymore. We will discuss some of these new phenomena in this chapter.

2.1 Metric Spaces

Before defining general topological spaces we recall some basic definitions and results on metric spaces as they should be familiar from undergraduate courses. We start recalling the main definition of a metric space:

Definition 2.1.1 (*Metric space*) A metric space is a pair (M, d) of a (non-empty) set M together with a map

$$d: M \times M \longrightarrow [0, \infty), \qquad (2.1.1)$$

called the metric, such that

(i) $d(x, y) = 0$ iff $x = y$,
(ii) $d(x, y) = d(y, x)$,
(iii) $d(x, y) \le d(x, z) + d(z, y)$

© Springer International Publishing Switzerland 2014
S. Waldmann, *Topology*, DOI 10.1007/978-3-319-09680-3_2

for all $x, y, z \in M$.

If one relaxes the first condition to $d(x, y) = 0$ if $x = y$ then d is sometimes called a *pseudo-metric*. The third condition is the triangle inequality. The geometric idea behind this definition is that $d(x, y)$ is a measure for the *distance* between the points x and y.

There are some easy and well-known examples of metric spaces; the verification of the defining properties is a simple exercise.

Example 2.1.2 (Metric spaces)

(i) The real numbers $\mathbb{R}$ with the usual absolute value $|\cdot|$ give a metric $d(x, y) = |x - y|$. Analogously, $\mathbb{C}$ becomes a metric space, too.

(ii) The previous example can be generalized in various ways. One important way is a real or complex normed vector space $(V, \| \cdot \|)$. Then $d(x, y) = \|x - y\|$ defines a metric on V. Here we mention the following particular cases:

- $\mathbb{R}^n$ or $\mathbb{C}^n$ with the Euclidean metric

$$d(x, y) = \sqrt{\sum_{k=1}^{n} |x_k - y_k|^2}. \tag{2.1.2}$$

- $\mathbb{R}^n$ or $\mathbb{C}^n$ with the metric coming from the *p-norm* with $p \in [1, \infty)$,

$$d_p(x, y) = \sqrt[p]{\sum_{k=1}^{n} |x_k - y_k|^p}. \tag{2.1.3}$$

- $\mathbb{R}^n$ or $\mathbb{C}^n$ with the metric coming from the supremum norm

$$d_\infty(x, y) = \|x - y\|_\infty = \max_{k=1}^{n} |x_k - y_k|. \tag{2.1.4}$$

(iii) Consider the space $\mathbb{R}[[x]]$ of real *formal power series* $a = \sum_{n=0}^{\infty} a_n x^n$ with coefficients $a_n \in \mathbb{R}$ and a variable x, where we do not care about convergence at all. By termwise operations $\mathbb{R}[[x]]$ is a real vector space. We define now the *order* of a to be

$$o(a) = \begin{cases} \infty & \text{if } a = 0 \\ \min\{n \mid a_n \neq 0\} & \text{if } a \neq 0. \end{cases} \tag{2.1.5}$$

Then the definition

$$d(a, b) = 2^{-o(a-b)} \tag{2.1.6}$$

gives a metric on $\mathbb{R}[[x]]$ where we set $2^{-\infty} = 0$ as usual, see also Exercise 2.7.2 for more details on this metric space.

(iv) If M is any set we define the *discrete metric* by

$$d(p, q) = \begin{cases} 0 & \text{if } p = q \\ 1 & \text{if } p \neq q. \end{cases} \qquad (2.1.7)$$

An easy verification shows that this is indeed a metric space with $d(p, q) \leq 1$ for all points $p, q \in M$.

(v) If (M, d_M) is a metric space and $N \subseteq M$ a subset then $d_N = d|_{N \times N}$ gives a metric on N, called the *induced metric* or *subspace metric*. This way, many interesting geometric objects like the spheres

$$\mathbb{S}^n = \left\{ x \in \mathbb{R}^{n+1} \ \middle| \ \|x\| = 1 \right\} \qquad (2.1.8)$$

or the tori

$$\mathbb{T}^n = \left\{ z \in \mathbb{C}^n \ \middle| \ z = (z_1, \ldots, z_n) \ \text{ with } \ |z_1| = \cdots = |z_n| = 1 \right\} \qquad (2.1.9)$$

inherit a metric structure by considering them as subsets of a suitably chosen Euclidean space.

(vi) If (M, d) is a metric space then also

$$d'(p, q) = \frac{d(p, q)}{1 + d(p, q)} \qquad (2.1.10)$$

is a metric on M. Now all points have distance $d'(p, q) < 1$. To validate the triangle inequality for d' we first note that the function $f : \xi \mapsto \frac{\xi}{1+\xi}$ is monotonically increasing on $\mathbb{R}_0^+$ and we have $f(\alpha + \beta) \leq f(\alpha) + f(\beta)$ for all $\alpha, \beta \geq 0$.

We see that there is a whole world of metric spaces to be explored. As a first step we define the open subsets of a metric space in a similar way open subsets of $\mathbb{R}$ are defined in elementary calculus.

Definition 2.1.3 (*Open subsets in metric space*) Let (M, d) be a metric space.

(i) The open ball $\mathrm{B}_r(p)$ around $p \in M$ of radius $r > 0$ is defined by

$$\mathrm{B}_r(p) = \{q \in M \mid d(p, q) < r\}. \qquad (2.1.11)$$

(ii) A subset $\mathcal{O} \subseteq M$ is called open if for all $p \in \mathcal{O}$ one finds a radius $r > 0$ (depending on p) such that $\mathrm{B}_r(p) \subseteq \mathcal{O}$.

Fig. 2.1 Open metric balls
are open

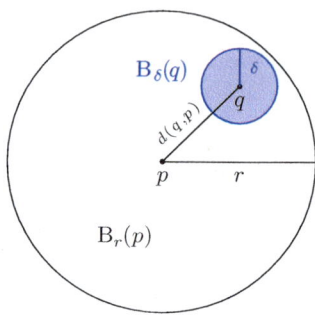

Of course, calling a ball open does not make it an open subset in the sense of
the definition. Here we have to prove something. We collect this and some further
properties of open subsets in the following proposition:

Proposition 2.1.4 (Open subsets in metric space) *Let (M, d) be a metric space.*

(i) All open balls $B_r(p)$ with $p \in M$ and $r > 0$ are open.
(ii) The empty set $\emptyset$ and M are open.
*(iii) If $\{\mathcal{O}_i\}_{i \in I}$ is an arbitrary collection of open subsets of M then $\bigcup_{i \in I} \mathcal{O}_i$ is open,
too.*
*(iv) If $\mathcal{O}_1, \dots, \mathcal{O}_n$ are finitely many open subsets of M then $\mathcal{O}_1 \cap \dots \cap \mathcal{O}_n$ is open,
too.*

Proof Let $q \in B_r(p)$ be given. Then $r - d(p, q) > 0$ and thus we find a $\delta > 0$ with
$\delta < r - d(p, q)$, see Fig. 2.1. Now consider $q' \in B_\delta(q)$. We claim that $q' \in B_r(p)$.
Indeed, by the triangle inequality we have

$$d(p, q') \le d(p, q) + d(q, q') < d(p, q) + \delta < r.$$

Thus $B_\delta(q) \subseteq B_r(p)$, showing that $B_r(p)$ is open. The second part is clear. For the
third, let $p \in \bigcup_{i \in I} \mathcal{O}_i$ then there is at least one $i_0 \in I$ with $p \in \mathcal{O}_{i_0}$. Since $\mathcal{O}_{i_0}$ is
open, we find $r > 0$ with $B_r(p) \subseteq \mathcal{O}_{i_0}$. But then

$$B_r(p) \subseteq \mathcal{O}_{i_0} \subseteq \bigcup_{i \in I} \mathcal{O}_i$$

shows that the union is open. For the last part, let $p \in \mathcal{O}_1 \cap \dots \cap \mathcal{O}_n$ if the intersection
is non-empty. Then we have $r_1, \dots, r_n > 0$ with $B_{r_1}(p) \subseteq \mathcal{O}_1, \dots, B_{r_n}(p) \subseteq \mathcal{O}_n$
since all of them are open. Taking the minimum $r = \min\{r_1, \dots, r_n\}$ gives an open
ball $B_r(p)$ contained in all $B_{r_1}(p), \dots, B_{r_n}(p)$. Thus $B_r(p) \subseteq \mathcal{O}_1 \cap \dots \cap \mathcal{O}_n$ follows,
showing the last part. $\qquad\square$

Analogously to the open balls we define the *closed balls* to be

$$B_r(p)^{\mathrm{cl}} = \{q \in M \mid d(p, q) \le r\}, \tag{2.1.12}$$

where again $p \in M$ and $r > 0$. Then we have the trivial inclusion

$$B_r(p) \subseteq B_r(p)^{\mathrm{cl}} \tag{2.1.13}$$

for all $p \in M$ and all $r > 0$. Moreover, a subset $A \subseteq M$ is called *closed* if its complement $M \setminus A$ is open. We get the following characterization of closed subsets:

Proposition 2.1.5 *Let (M, d) be a metric space.*

(i) All closed balls are closed.
(ii) The empty set $\emptyset$ and M are closed.
(iii) If $\{A_i\}_{i \in I}$ is an arbitrary collection of closed subsets of M then $\bigcap_{i \in I} A_i$ is closed, too.
(iv) If $A_1, \ldots, A_n$ are finitely many closed subsets then their union $A_1 \cup \cdots \cup A_n$ is closed, too.

Proof Let $B_r(p)^{\mathrm{cl}}$ be given and $q \in M \setminus B_r(p)^{\mathrm{cl}}$. Then $d(p, q) > r$ and hence we find a $\delta > 0$ with $r + \delta < d(p, q)$. Now let $q' \in B_\delta(q)$. Then by the triangle inequality

$$d(p, q) \le d(p, q') + d(q', q) < d(p, q') + \delta,$$

and thus $d(p, q') > d(p, q) - \delta > r$ showing that $q' \in M \setminus B_r(p)^{\mathrm{cl}}$. This gives the first part. The remaining parts are obtained from the corresponding statements on open subsets in Proposition 2.1.4 by passing to the complements. □

In a next step we recall the $\epsilon\delta$-definition of continuity and translate it into a statement involving only the open subsets. This is quite remarkable as the open subsets carry much less information than the metric:

Example 2.1.6 Let (M, d) be a metric space and define the metric d' as in Example 2.1.2, (vi). Then $\mathcal{O} \subseteq M$ is open with respect to d iff it is open with respect to d'. Indeed, the function $f(\xi) = \frac{\xi}{1+\xi}$ for $\xi \in [0, \infty)$ has the inverse function $f^{-1}(\eta) = \frac{\eta}{1-\eta}$. From this it follows that $d(p, q) < r$ iff $d'(p, q) < r'$ where $r' = f(r)$. This shows that the open balls for the two metrics coincide after rescaling the radii with the maps f and f^{-1}, respectively.

Definition 2.1.7 *($\epsilon\delta$-Continuity)* Let $f : (M, d_M) \longrightarrow (N, d_N)$ be a map between metric spaces.

(i) The map f is called continuous at $p \in M$ if for all $\varepsilon > 0$ there is a $\delta > 0$ with

$$d_N(f(p), f(q)) < \varepsilon \quad \text{for} \quad d_M(p, q) < \delta. \tag{2.1.14}$$

(ii) The map f is called continuous if f is continuous at all points $p \in M$.

Proposition 2.1.8 *Let $f : (M, d_M) \longrightarrow (N, d_N)$ be a map between metric spaces.*

(i) *The map f is continuous at $p \in M$ iff for every open subset $U \subseteq N$ with $f(p) \in U$ the preimage $f^{-1}(U)$ contains an open subset $\mathcal{O} \subseteq f^{-1}(U)$ with $p \in \mathcal{O}$.*

(ii) *The map f is continuous iff the preimage $f^{-1}(U)$ of every open subset U of N is open in M.*

Proof Let f be continuous at p and let $U \subseteq N$ be open with $f(p) \in U$. Then there exists an open ball $\mathrm{B}_\varepsilon(f(p)) \subseteq U$ for some $\varepsilon > 0$. By assumption, we find a $\delta > 0$ for which (2.1.14) applies. Thus for $q \in \mathrm{B}_\delta(p)$ we have $f(q) \in \mathrm{B}_\varepsilon(f(p)) \subseteq U$ showing that $\mathrm{B}_\delta(p) \subseteq f^{-1}(U)$. Conversely, assume f fulfills the condition of (i) and let $\varepsilon > 0$ be given. Then $f^{-1}(\mathrm{B}_\varepsilon(f(p)))$ contains an open subset $\mathcal{O}$ which contains p. But then there is a $\delta > 0$ with $\mathrm{B}_\delta(p) \subseteq \mathcal{O} \subseteq f^{-1}(\mathrm{B}_\varepsilon(f(p)))$. This means that for $q \in M$ with $d_M(p, q) < \delta$ we have $f(q) \in \mathrm{B}_\varepsilon(f(p))$ which gives $d_N(f(p), f(q)) < \varepsilon$. Thus the first part is shown. For the second assume first that f is continuous and $\mathcal{O} \subseteq N$ is open. We can now apply the first part to every $p \in f^{-1}(\mathcal{O})$, to show that there is an open ball around p which is entirely contained in $f^{-1}(\mathcal{O})$. Thus $f^{-1}(\mathcal{O})$ is open. Conversely, assume $f^{-1}(\mathcal{O})$ is open for every open $\mathcal{O}$. Again, we can apply the first part to every $p \in M$ and every open ball $\mathrm{B}_\varepsilon(f(p))$ and get the openness of $f^{-1}(\mathrm{B}_\varepsilon(f(p)))$. Thus there is an open ball $\mathrm{B}_\delta(p) \subseteq f^{-1}(\mathrm{B}_\varepsilon(f(p)))$ which gives the continuity at p. $\square$

Passing again to complements we get the following alternative characterization using closed subsets:

Corollary 2.1.9 *Let $f : (M, d_M) \longrightarrow (N, d_N)$ be a map between metric spaces. Then f is continuous iff the preimage of every closed subset of N is closed in M.*

In the formulation of the continuity at a given point the subsets which contain an open ball $\mathrm{B}_r(p)$ play a particular role. This justifies the following definition:

Definition 2.1.10 (*Neighbourhood*) Let (M, d) be a metric space and let $p \in M$. Then a subset $U \subseteq M$ is called neighbourhood of p if $p \in U$ and U contains an open ball $\mathrm{B}_r(p) \subseteq U$. The set of all neighbourhoods of p is denoted by $\mathfrak{U}(p)$.

Note that a neighbourhood U of p needs not to be open itself, though it always contains an open neighbourhood. The following proposition lists a few elementary properties of neighbourhoods:

Proposition 2.1.11 *Let (M, d) be a metric space and let $p \in M$.*

(i)   *If U is a neighbourhood of p and $U \subseteq U'$ then U' is a neighbourhood of p, too.*

(ii) *If $U_1, \ldots, U_n$ are neighbourhoods of p then $U_1 \cap \cdots \cap U_n$ is a neighbourhood of p, too.*

(iii) *Any neighbourhood of p contains p.*

(iv) If U is a neighbourhood of p then there exists a neighbourhood $V \subseteq U$ of p
    such that V is a neighbourhood of all $q \in V$.

Proof The first part is clear since with $B_r(p) \subseteq U$ we have also $B_r(p) \subseteq U'$. The second part is obtained by taking again the minimum of the relevant radii. The third is clear. For the fourth part we note that an open subset, like an open ball, is a neighbourhood of all its points. This was precisely the content the definition of an open set in Definition 2.1.3, (ii). Then the fourth part follows by taking $V = B_r(p) \subseteq U$ for a suitable radius $r > 0$.   □

Using the notion of neighbourhoods we can rephrase the statement of Proposition 2.1.8, (i), as follows:

Corollary 2.1.12 *A map $f: (M, d_M) \longrightarrow (N, d_N)$ between metric spaces is continuous at p if the preimage of every neighbourhood of $f(p)$ is a neighbourhood of p.*

Again we see that the notion of a neighbourhood refers to the open subsets of M only: the detailed information about the metric is also lost here, the neighbourhood systems of the metric spaces (M, d) and (M, d') with d' as in Example 2.1.2, (vi), are the same even though the metrics are very different.

The next concept is also transferred easily from elementary calculus to metric spaces: convergence of sequences and the notion of completeness.

Definition 2.1.13 *(Convergence and completeness)* Let (M, d) be a metric space and let $(p_n)_{n \in \mathbb{N}}$ be a sequence in M.

(i) The sequence $(p_n)_{n \in \mathbb{N}}$ converges to $p \in M$ if for every $\varepsilon > 0$ one finds an $N \in \mathbb{N}$ such that for all $n \geq N$ one has

$$d(p, p_n) < \varepsilon. \tag{2.1.15}$$

(ii) The sequence $(p_n)_{n \in \mathbb{N}}$ is called a Cauchy sequence if for all $\varepsilon > 0$ one finds an $N \in \mathbb{N}$ such that for all $n, m \geq N$ one has

$$d(p_n, p_m) < \varepsilon. \tag{2.1.16}$$

(iii) The metric space (M, d) is called complete if every Cauchy sequence converges.

The third part is reasonable as every convergent sequence is clearly a Cauchy sequence by the triangle inequality. Again we can formulate convergence in form of open subsets and neighbourhoods alone. The notion of Cauchy sequences and completeness requires some additional structure which we shall not consider at the moment.

Proposition 2.1.14 *Let (M, d) be a metric space and let $(p_n)_{n \in \mathbb{N}}$ be a sequence in M. Then $(p_n)_{n \in \mathbb{N}}$ converges to $p \in M$ iff for every neighbourhood U of p there is an $N \in \mathbb{N}$ such that for all $n \geq N$ we have $p_n \in U$.*

Proof The equivalence of the two statements is now easy to see, we leave this as a little task for the reader.   □

As a conclusion of this discussion we arrive at the point of view that many features of metric spaces actually do not depend on the metric but only on the system of open subsets. This suggest to consider an axiomatized version of "open subsets" from the beginning and study this theory of topological spaces instead of the theory of metric spaces. It will turn out that this has several benefits:

(i) There will be important examples of "spaces" which can not be treated as metric spaces but as topological spaces only: among many other examples the notion of a differentiable manifold in differential geometry does not refer to a metric space from the beginning but to a topological space (even though it turns out a posteriori that they do carry a compatible metric structure). In functional analysis many important topological vector spaces are known to be "non-metrizable" like e.g. the space of test functions $\mathscr{C}_0^\infty(\mathbb{R})$. A precise formulation and a proof of this statement will require some more advanced technology which we will learn in the sequel, see Exercise 7.4.8.
(ii) Many concepts and proofs simplify drastically after taking the point of view of topological spaces compared to the usage of metric spaces, even though they also apply for metric spaces.
(iii) Topology will be an ideal playground to practice "axiomatization" of a mathematical concept, a technology which will be useful at many other places as well.

On the other hand one should not forget that the theory of metric spaces provides a finer structure and thus more specific features which will not be captured by topological spaces, the notion of Cauchy sequences and completeness is one first example. There is an appropriate axiomatization of this as well in the theory of uniform spaces, but for the time being we shall not touch this. Another important aspect of metric spaces will be the theory of coarse spaces or asymptotic geometry, certainly asking for a separate course.

2.2 Topological Spaces

We come now to the definition of a general topological space. There are at least two approaches: Historically, Hausdorff axiomatized the properties of the neighbourhoods of a point as found in [10, Chap. XII, §1]. Shortly later, Alexandroff axiomatized the properties of the system of open subsets, which is now the usual approach. In any case, both definitions turn out to be equivalent. For a given set M we denote the power set of M by 2^M.

The main idea is now very simple: we want to axiomatize the behaviour of open subsets in a metric space as found in Proposition 2.1.4 without explicit reference to the metric itself:

Definition 2.2.1 (*Topological space*) Let M be a set. Then a subset $\mathcal{M} \subseteq 2^M$ of the power set of M is called a topology if the following properties are fulfilled:

(i) The empty set $\emptyset$ and M are in $\mathcal{M}$.

(ii) If $\{\mathcal{O}_i\}_{i \in I}$ with $\mathcal{O}_i \in \mathcal{M}$ is an arbitrary collection of elements in $\mathcal{M}$ then also $\bigcup_{i \in I} \mathcal{O}_i \in \mathcal{M}$.

(iii) If $\mathcal{O}_1, \ldots, \mathcal{O}_n \in \mathcal{M}$ are finitely many elements in $\mathcal{M}$ then also $\mathcal{O}_1 \cap \cdots \cap \mathcal{O}_n \in \mathcal{M}$.

A set M together with a topology $\mathcal{M} \subseteq 2^M$ is called a topological space $(M, \mathcal{M})$ and the sets $\mathcal{O} \in \mathcal{M}$ are called the open subsets of $(M, \mathcal{M})$.

Other commonly used notations for topologies on M are τ_M or T_M, sometimes also just τ if the reference to M is clear.

Example 2.2.2 (Topologies) Let M be a set.

(i) The power set $\mathcal{M}_{\text{discrete}} = 2^M$ is a topology on M, called the *discrete* or *finest topology* of M.

(ii) Taking $\mathcal{M}_{\text{indiscrete}} = \{\emptyset, M\}$ gives a topology, called the *indiscrete* or *trivial* or *coarsest topology* of M.

(iii) If d is a metric for M then taking $\mathcal{M}$ to be the open subsets in the sense of Definition 2.1.3 gives a topology on M, called the *metric topology* of the metric space (M, d). From e.g. Example 2.1.6 and Exercise 2.7.1 we see that very different metrics can yield the same topology. Note also that the metric topology of the discrete metric on a set M from Example 2.1.2, (iv), is the discrete topology on M.

(iv) Consider the collection $\mathcal{M}_{\text{cofinite}} \subseteq 2^M$ of all those subsets which have *finite* complements and $\emptyset$. This yields the *cofinite topology* on M.

Complementary to the definition of open subsets we have the closed subsets:

Definition 2.2.3 (*Closed subset*) A subset $A \subseteq M$ of a topological space $(M, \mathcal{M})$ is called closed if $M \setminus A$ is open.

As in Proposition 2.1.5 we concluded that $\emptyset$ and M are both closed and finite unions as well as arbitrary intersections of closed subsets are again closed. Thus we can equivalently characterize a topological space by means of its closed subsets instead of the open ones.

The next observation will be important for constructing topologies:

Proposition 2.2.4 *If $\{\mathcal{M}_i\}_{i \in I}$ are topologies on M then also $\mathcal{M} = \bigcap_{i \in I} \mathcal{M}_i$ is a topology on M.*

Proof Note that here we take the intersection of subsets of the power set 2^M. Let $\mathcal{M} = \bigcap_{i \in I} \mathcal{M}_i \subseteq 2^M$. Since $\emptyset, M \in \mathcal{M}_i$ for all $i \in I$ we also have $\emptyset, M \in \mathcal{M}$. Now let $\{\mathcal{O}_j\}_{j \in J}$ be a collection of subsets in $\mathcal{M}$, i.e. $\mathcal{O}_j \in \mathcal{M}$. Then $\mathcal{O}_j \in \mathcal{M}_i$ for all $i \in I$ and hence also $\bigcup_{j \in J} \mathcal{O}_j \in \mathcal{M}_i$ for all $i \in I$, since $\mathcal{M}_i$ is a topology. But this means $\bigcup_{j \in J} \mathcal{O}_j \in \mathcal{M}$, too. Analogously, we get for $\mathcal{O}_1, \ldots, \mathcal{O}_n \in \mathcal{M}$ that $\mathcal{O}_1 \cap \cdots \cap \mathcal{O}_n \in \mathcal{M}$. $\qquad \square$

Given two topologies $\mathcal{M}_1$, $\mathcal{M}_2$ on M we want to compare them. Since $\mathcal{M}_1$, $\mathcal{M}_2 \in 2^M$ are both subsets of a common set, the power set, there is an obvious way of how to do that:

Definition 2.2.5 (*Finer and coarser*) Let $\mathcal{M}_1$, $\mathcal{M}_2 \subseteq 2^M$ be topologies on M. Then $\mathcal{M}_1$ is called finer than $\mathcal{M}_2$ if $\mathcal{M}_2 \subseteq \mathcal{M}_1$. In this case $\mathcal{M}_2$ is called coarser than $\mathcal{M}_1$.

This explains the notation in Example 2.2.2, (i) and (ii): the discrete topology is indeed the finest possible topology on M we can have while the indiscrete topology is the coarsest one. Note that we use "finer" in the sense of "finer or equal" but not in the sense of "strictly finer".

Often we have a space M together with a collection $\mathcal{S} \subseteq 2^M$ of subsets which we would like to be the open subsets of a topology. However, $\mathcal{S}$ may fail to be a topology directly. Thus one is looking for a topology containing $\mathcal{S}$ in a economical way: it should be as coarse as possible. Clearly the finest topology $\mathcal{M}_{\mathrm{finest}} = 2^M$ will contain $\mathcal{S}$ for trivial reasons so this other extreme case is uninteresting.

Proposition 2.2.6 *Let $\mathcal{S} \subseteq 2^M$ be a subset of the power set of M containing $\emptyset$ and M.*

(i) *There exists a unique topology $\mathcal{M}(\mathcal{S})$ which is coarser than every other topology containing $\mathcal{S}$.*

(ii) *This topology $\mathcal{M}(\mathcal{S})$ can be obtained by the following two-step procedure: first we take all finite intersections of subsets in $\mathcal{S}$ and afterwards we take arbitrary unions of the resulting subsets.*

Proof Since $\mathcal{M}_{\mathrm{discrete}} = 2^M$ is a topology containing $\mathcal{S}$ there is at least one topology containing $\mathcal{S}$. Now we take the intersection of *all* these topologies containing $\mathcal{S}$. By Proposition 2.2.4 this is again a topology, which still contains $\mathcal{S}$. Apparently, it is contained in any other topology containing $\mathcal{S}$ by the very construction. Thus it is also the unique one with this property, proving the first part. For the second part, let $\tilde{\mathcal{M}}(\mathcal{S})$ be the collection of subsets we get by first taking finite intersections and arbitrary unions afterwards. Clearly, $\tilde{\mathcal{M}}(\mathcal{S}) \subseteq \mathcal{M}(\mathcal{S})$ since a topology is stable under taking finite intersections and arbitrary unions. We have to show that they are equal. If $\mathcal{O}_i \in \tilde{\mathcal{M}}(\mathcal{S})$ for $i \in I$ with some index set I then each $\mathcal{O}_i = \bigcup_{j \in J_i} \mathcal{O}_{ij}$ and each $\mathcal{O}_{ij} = S_{ij1} \cap \cdots \cap S_{ijn}$ with $S_{ij1}, \ldots, S_{ijn} \in \mathcal{S}$ and n depending on i and j. Note that by repeating $\mathcal{O}_{ij}$'s we can assume that the index set $J = J_i$ is actually the same for all i. But then

$$\mathcal{O} = \bigcup_{i \in I} \mathcal{O}_i = \bigcup_{i \in I, j \in J} \mathcal{O}_{ij} = \bigcup_{i \in I, j \in J} S_{ij1} \cap \cdots \cap S_{ijn}$$

is again a union of the desired form. Hence $\mathcal{O} \in \tilde{\mathcal{M}}(\mathcal{S})$. For a finite I we get

$$\bigcap_{i \in I} \mathcal{O}_i = \bigcap_{i \in I} \bigcup_{j \in J} \mathcal{O}_{ij} = \bigcup_{j \in J} \bigcap_{i \in I} S_{ij1} \cap \cdots \cap S_{ijn}.$$

But the intersection is always finite for a fixed $j \in J$. Hence also $\mathcal{O} \in \tilde{\mathcal{M}}(\mathcal{S})$ proving that $\tilde{\mathcal{M}}(\mathcal{S})$ is a topology. By the first part it has to coincide with $\mathcal{M}(\mathcal{S})$. $\square$

Of course the condition $\emptyset, M \in \mathcal{S}$ is easy to achieve by augmenting $\mathcal{S}$ if needed: we included this more by convenience. The construction in Proposition 2.2.4 allows to generate topologies very easily:

Example 2.2.7 The metric topology on a metric space (M, d) is the topology generated by the open balls, i.e. the coarsest topology containing all open balls. Indeed, an open subset $\mathcal{O} \subseteq M$ in the metric topology can be written as

$$\mathcal{O} = \bigcup_{p \in \mathcal{O}} B_{r(p)}(p) \tag{2.2.1}$$

with suitable radii $r(p) > 0$ such that $B_{r(p)}(p) \subseteq \mathcal{O}$. Thus here we do not even need to take finite intersections first. By Proposition 2.1.4 this is a topology, coarser than the one constructed in the second part of Proposition 2.2.6. By the first part of Proposition 2.2.6 it coincides with the coarsest one containing the open balls.

The subset $\mathcal{S}$ generating a topology is often very useful and justifies its own terminology:

Definition 2.2.8 (*Basis and Subbasis*) Let $(M, \mathcal{M})$ be a topological space and let $\mathcal{S}, \mathcal{B} \subseteq \mathcal{M}$ be subsets (containing already $\emptyset$ and M).

(i) The set $\mathcal{B}$ is called a basis of $\mathcal{M}$ if every open subset is a union of subsets from $\mathcal{B}$.
(ii) The set $\mathcal{S}$ is called a subbasis of $\mathcal{M}$ if the collection of all finite intersections of sets from $\mathcal{S}$ forms a basis of $\mathcal{M}$.

Clearly, a basis is also a subbasis and $\mathcal{S}$ is a subbasis if $\mathcal{M}(\mathcal{S}) = \mathcal{M}$. Thus, in view of Proposition 2.2.6, a subbasis seems to be the more important concept. For a metric space, Example 2.2.7 shows that the open balls form a basis and not just a subbasis: we do not need to take finite intersections of open balls, they are already obtained by taking suitable unions of smaller balls.

The last general construction we shall discuss here is given by the subspace topology. If $N \subseteq M$ is a subset of a topological space $(M, \mathcal{M})$ then we define

$$\mathcal{M}\big|_N = \{\mathcal{O} \cap N \mid \mathcal{O} \in \mathcal{M}\} \subseteq 2^N, \tag{2.2.2}$$

i.e. all subsets of N which are obtained by intersecting N with an open subset of M, see also Fig. 2.2.

Proposition 2.2.9 (Subspace topology) *Let $(M, \mathcal{M})$ be a topological space and $N \subseteq M$ a subset. Then $\mathcal{M}\big|_N$ is a topology for N.*

The verification is straightforward. Nevertheless, some caution is necessary: an open subset $U \subseteq N$ with respect to $\mathcal{M}\big|_N$ can also be considered as subset of M but then it fails to be open with respect to $\mathcal{M}$ in general. The same holds for closed subsets of $(N, \mathcal{M}\big|_N)$ viewed as subsets of $(M, \mathcal{M})$, see also Exercise 2.7.3.

Fig. 2.2 An open set in the subspace topology

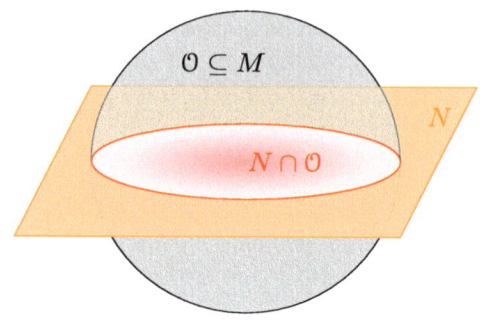

2.3 Neighbourhoods, Interiors, and Closures

As in the metric case we define now the neighbourhood system of a point in a topological space:

Definition 2.3.1 (*Neighbourhood*) Let $(M, \mathcal{M})$ be a topological space and $p \in M$.

(i) A subset $U \subseteq M$ is called neighbourhood of p if there exists an open subset $\mathcal{O} \subseteq M$ with $p \in \mathcal{O} \subseteq U$.
(ii) The collection of all neighbourhoods of p is called the neighbourhood system (or neighbourhood filter) of p, denoted by $\mathfrak{U}(p)$.

The properties of neighbourhoods in the metric case as discussed in Proposition 2.1.11 carry over to the general situation of a topological space:

Proposition 2.3.2 *Let $(M, \mathcal{M})$ be a topological space and $p \in M$.*

(i) A subset $\mathcal{O} \subseteq M$ is a neighbourhood of all of its points iff $\mathcal{O}$ is open.
(ii) For $U \in \mathfrak{U}(p)$ and $U \subseteq U'$ we have $U' \in \mathfrak{U}(p)$.
(iii) For $U_1, \ldots, U_n \in \mathfrak{U}(p)$ we have $U_1 \cap \cdots \cap U_n \in \mathfrak{U}(p)$.
(iv) For $U \in \mathfrak{U}(p)$ we have $p \in U$.
(v) For $U \in \mathfrak{U}(p)$ there exists a $V \in \mathfrak{U}(p)$ with $V \subseteq U$ and $V \in \mathfrak{U}(q)$ for all $q \in V$.

Proof Suppose $\mathcal{O}$ is open, then clearly $\mathcal{O}$ is a neighbourhood for all $p \in \mathcal{O}$. Thus assume $\mathcal{O} \in \mathfrak{U}(p)$ for all $p \in \mathcal{O}$. Then we get an open $\mathcal{O}_p \subseteq M$ for every $p \in \mathcal{O}$ with $p \in \mathcal{O}_p \subseteq \mathcal{O}$. It follows that $\mathcal{O} = \bigcup_{p \in \mathcal{O}} \mathcal{O}_p$ is open, proving the first part. The parts (ii)–(v) are now analogous to the statements of Proposition 2.1.11. $\quad\square$

Remark 2.3.3 Conversely, given a non-empty system of subsets $\mathfrak{U}(p)$ for every $p \in M$ satisfying the properties (ii)–(v), which we then call a *system of neighbourhoods*, we can reconstruct a unique topology $\mathcal{M}$ on M such that the $\mathfrak{U}(p)$ are the neighbourhood systems with respect to $\mathcal{M}$: One defines $\mathcal{O} \subseteq M$ to be open if it is a neighbourhood of all of its points. This was Hausdorff's original approach in [10, Sect. Chap. VII, §1], see also Exercise 2.7.4.

Definition 2.3.4 (*Neighbourhood basis*) Let $(M, \mathcal{M})$ be a topological space and $p \in M$. Then a subset $\mathfrak{B}(p) \subseteq \mathfrak{U}(p)$ of neighbourhoods of $p \in M$ is called a neighbourhood basis of p if for every $U \in \mathfrak{U}(p)$ there is a $B \in \mathfrak{B}(p)$ with $B \subseteq U$.

Example 2.3.5 (Neighbourhood bases) Let M be a set.

(i) For a metric d on M and $p \in M$ the open balls $\mathrm{B}_{r_n}(p)$ with $r_n > 0$ being a zero sequence constitute a *countable* neighbourhood basis for the metric topology.
(ii) For the discrete topology $\mathcal{M}_{\mathrm{discrete}}$ on M the set $\{\{p\}\} = \mathfrak{B}(p)$ is a *finite* neighbourhood basis since $\{p\}$ is open.
(iii) Let M be uncountable and consider the cofinite topology $\mathcal{M}_{\mathrm{cofinite}}$. Then a neighbourhood basis of $p \in M$ is necessarily uncountable. While this seems to be a rather artificial example there are (more non-trivial) examples of function spaces in functional analysis which have topologies such that neighbourhood bases are very large, i.e. uncountable.

To capture this phenomenon of having very different sizes of neighbourhood bases, one introduces the following two countability axioms:

Definition 2.3.6 (*First and second countability*) Let $(M, \mathcal{M})$ be a topological space.

(i) The space M is called first countable (at $p \in M$) if every point (the point p) has a countable neighbourhood basis.
(ii) The space M is called second countable if $\mathcal{M}$ has a countable basis.

Clearly second countable implies first countable, the converse needs not to be true. The second countability plays an important role in e.g. differential geometry where it guarantees that manifolds do not become "too big".

Example 2.3.7 The metric topology of $\mathbb{R}^n$ is second countable as it is sufficient to consider only those open balls $\mathrm{B}_r(x)$ where $r > 0$ is rational and $x \in \mathbb{Q}^n$, see Exercise 2.7.6.

Given an arbitrary subset $A \subseteq M$ of a topological space $(M, \mathcal{M})$ one can construct certain other subsets using the topology. The following definitions are motivated by the geometric intuition in $\mathbb{R}^n$:

Definition 2.3.8 (*Interior, closure and boundary*) Let $(M, \mathcal{M})$ be a topological space and $A \subseteq M$.

(i) A point $p \in M$ is called inner point of A if $A \in \mathfrak{U}(p)$.
(ii) The interior A° of A is the set of all inner points of A.
(iii) A point $p \in M$ is called boundary point of A if for every neighbourhood $U \in \mathfrak{U}(p)$ we have $A \cap U \neq \emptyset \neq (M \setminus A) \cap U$.
(iv) The boundary ∂A of A is the set of all boundary points of A.
(v) The closure A^{cl} of A is the set of all points $p \in M$ such that all $U \in \mathfrak{U}(p)$ satisfy $U \cap A \neq \emptyset$.

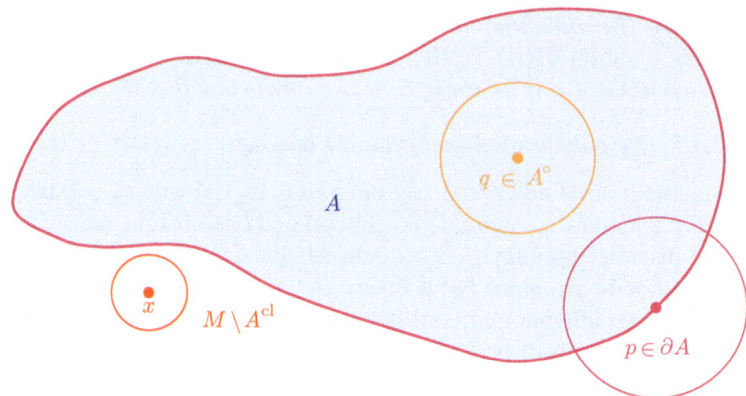

Fig. 2.3 The open interior, the boundary and the closure of a subset A: one has $q \in A^\circ$, $p \in \partial A$, and $x \in M \setminus A^{\mathrm{cl}}$

A commonly used alternative notation is $\overline{A}$ for the closure A^{cl} and $\mathrm{int}(A)$ for the interior A°. The geometric intuition behind these definitions can easily be visualized, see Fig. 2.3. We get the following alternative characterizations of the interior, the closure, and the boundary.

Proposition 2.3.9 *Let $(M, \mathcal{M})$ be a topological space and $A \subseteq M$.*

(i) The interior A° of A is the largest open subset inside A.
(ii) The closure A^{cl} of A is the smallest closed subset containing A.
(iii) The boundary ∂A of A is closed and $\partial A = A^{\mathrm{cl}} \setminus A^\circ$.

Proof First we show that A° is open at all. For $p \in A^\circ$ we know $A \in \mathfrak{U}(p)$ and hence there is an open subset $U \subseteq A$ with $p \in U$. But then for all $q \in U$ we have $A \in \mathfrak{U}(q)$ as well, showing $q \in A^\circ$, too. Hence $U \subseteq A^\circ$ follows, i.e. $A^\circ \in \mathfrak{U}(p)$. Since this holds for all $p \in A^\circ$, we have an open subset A° by Proposition 2.3.2, (i). Conversely, if $\mathcal{O} \subseteq A$ is an open subset then $p \in \mathcal{O} \subseteq A$ is an interior point of A, showing $\mathcal{O} \subseteq A^\circ$. Thus the first statement follows. For the second part, let $p \notin A^{\mathrm{cl}}$ then there exists a $U \in \mathfrak{U}(p)$ with $U \cap A = \emptyset$. Without restriction, we can assume U to be open. But then $U \in \mathfrak{U}(q)$ for all $q \in U$ showing that also $q \notin A^{\mathrm{cl}}$. Hence $U \subseteq M \setminus A^{\mathrm{cl}}$ implies that for every point $p \in M \setminus A^{\mathrm{cl}}$ a whole open neighbourhood of p is in $M \setminus A^{\mathrm{cl}}$. Thus $M \setminus A^{\mathrm{cl}}$ is open and A^{cl} is closed. Clearly $A \subseteq A^{\mathrm{cl}}$. Now let $\tilde{A} = \bigcap_{A \subseteq B, B \text{ closed}} B$ be the intersection of all closed subsets containing A. Clearly $\tilde{A}$ is the smallest closed subset containing A and thus $\tilde{A} \subseteq A^{\mathrm{cl}}$. Passing to complements gives

$$M \setminus \tilde{A} = \bigcup_{\substack{U \subseteq M \setminus A \\ U \text{ is open}}} U,$$

and hence for $p \in M \setminus \tilde{A}$ one has an open subset $U \subseteq M \setminus A$ with $p \in U$. But then $p \in U$ with $U \cap A = \emptyset$ shows $p \notin A^{\mathrm{cl}}$. Thus $A^{\mathrm{cl}} \cap (M \setminus \tilde{A}) = \emptyset$ or $A^{\mathrm{cl}} \subseteq \tilde{A}$ which proves (ii). For the last part we first note that $\partial A \subseteq A^{\mathrm{cl}}$. In fact, the definition shows

$$\partial A = A^{\mathrm{cl}} \cap (M \setminus A)^{\mathrm{cl}},$$

and hence ∂A is closed by (ii). Now let $p \in A^{\mathrm{cl}} \setminus \partial A$ then for all $U \in \mathfrak{U}(p)$ we have $U \cap A \neq \emptyset$ but there is a $U \in \mathfrak{U}(p)$ with $U \cap (M \setminus A) = \emptyset$. For this U we have $U \subseteq A$. This shows that every point $p \in A^{\mathrm{cl}} \setminus \partial A$ is inner, i.e. $A^{\mathrm{cl}} \setminus \partial A \subseteq A^{\circ}$. Conversely, for an interior point $p \in A^{\circ}$ we clearly have $p \notin \partial A$ but $p \in A^{\mathrm{cl}}$, showing the last part. □

By taking the closure, a subset is enlarged. Taking the open interior enlarges the complement, in both cases by the missing boundary points. The following two extreme cases will be of particular interest:

Definition 2.3.10 (*Dense and nowhere dense*) Let $(M, \mathcal{M})$ be a topological space.

(i) A subset $A \subseteq M$ is called dense if $A^{\mathrm{cl}} = M$.
(ii) A subset $A \subseteq M$ is called nowhere dense if $(A^{\mathrm{cl}})^{\circ} = \emptyset$.

Note that nowhere dense is strictly stronger than not dense, examples will be discussed in Exercise 2.7.24, see also Exercise 2.7.12.

We collect now some useful properties of closures, open interiors and boundaries and their behaviour with respect to unions, intersections, and complements.

Proposition 2.3.11 *Let $(M, \mathcal{M})$ be a topological space and let $A, B \subseteq M$ be subsets.*

(i) *One has $\emptyset^{\circ} = \emptyset = \emptyset^{\mathrm{cl}}$ and $M^{\circ} = M = M^{\mathrm{cl}}$ as well as $\partial \emptyset = \emptyset = \partial M$.*
(ii) *One has $A^{\circ} \subseteq A \subseteq A^{\mathrm{cl}}$ and*

$$(A^{\circ})^{\circ} = A^{\circ}, \quad \partial(\partial A) \subseteq \partial A, \quad and \quad (A^{\mathrm{cl}})^{\mathrm{cl}} = A^{\mathrm{cl}}. \tag{2.3.1}$$

(iii) *For $A \subseteq B$ one has $A^{\circ} \subseteq B^{\circ}$ and $A^{\mathrm{cl}} \subseteq B^{\mathrm{cl}}$.*
(iv) *One has*

$$A^{\circ} \cup B^{\circ} \subseteq (A \cup B)^{\circ}, \quad \partial(A \cup B) \subseteq \partial A \cup \partial B, \quad and \quad A^{\mathrm{cl}} \cup B^{\mathrm{cl}} = (A \cup B)^{\mathrm{cl}}. \tag{2.3.2}$$

(v) *One has*

$$A^{\circ} \cap B^{\circ} = (A \cap B)^{\circ} \quad and \quad (A \cap B)^{\mathrm{cl}} \subseteq A^{\mathrm{cl}} \cap B^{\mathrm{cl}}. \tag{2.3.3}$$

(vi) *One has*

$$(M \setminus A)^{\circ} = M \setminus A^{\mathrm{cl}}, \quad \partial(M \setminus A) = \partial A, \quad and \quad (M \setminus A)^{\mathrm{cl}} = M \setminus A^{\circ}. \tag{2.3.4}$$

Proof The first part is clear, either directly using the definitions or the characterizations from Proposition 2.3.9. For the second, an inner point of A is in particular a point in A and a point in A is in the closure since every neighbourhood $U \in \mathfrak{U}(p)$ intersects A at least in $\{p\}$ itself. Thus $A^\circ \subseteq A \subseteq A^{\mathrm{cl}}$ follows. Since A° is the largest open set inside A, we can apply this to A° and get that $(A^\circ)^\circ$ is the largest open set inside A°. Since A° is already open, we have $(A^\circ)^\circ = A^\circ$. Analogously, we can argue for the closure to get $(A^{\mathrm{cl}})^{\mathrm{cl}} = A^{\mathrm{cl}}$. Since a boundary is always closed, we have from Proposition 2.3.9, (iii), the relation $\partial(\partial A) = (\partial A)^{\mathrm{cl}} \setminus (\partial A)^\circ = \partial A \setminus (\partial A)^\circ \subseteq \partial A$, completing the second part. The third part is again clear from Proposition 2.3.9, (i) and (ii). For the fourth part we have $A^\circ \subseteq A \subseteq A \cup B$ and also $B^\circ \subseteq A \cup B$. Thus $A^\circ \cup B^\circ \subseteq A \cup B$ is an open subset of $A \cup B$ and hence contained in the largest such open subset, i.e. in $(A \cup B)^\circ$. Next, $A \subseteq A \cup B$ implies $A^{\mathrm{cl}} \subseteq (A \cup B)^{\mathrm{cl}}$ and analogously $B^{\mathrm{cl}} \subseteq (A \cup B)^{\mathrm{cl}}$ showing $A^{\mathrm{cl}} \cup B^{\mathrm{cl}} \subseteq (A \cup B)^{\mathrm{cl}}$. Hence $A^{\mathrm{cl}} \cup B^{\mathrm{cl}}$ is a closed subset containing $A \cup B$. Since the smallest such closed subset is $(A \cup B)^{\mathrm{cl}}$ we get $(A \cup B)^{\mathrm{cl}} \subseteq A^{\mathrm{cl}} \cup B^{\mathrm{cl}}$. Thus we have equality $A^{\mathrm{cl}} \cup B^{\mathrm{cl}} = (A \cup B)^{\mathrm{cl}}$. Finally, we get

$$\begin{aligned}
\partial(A \cup B) &= (A \cup B)^{\mathrm{cl}} \setminus (A \cup B)^\circ \\
&\subseteq (A^{\mathrm{cl}} \cup B^{\mathrm{cl}}) \setminus (A^\circ \cup B^\circ) \\
&\subseteq (A^{\mathrm{cl}} \setminus A^\circ) \cup (B^{\mathrm{cl}} \setminus B^\circ) \\
&= \partial A \cup \partial B,
\end{aligned}$$

since $A^\circ \subseteq A^{\mathrm{cl}}$ and $B^\circ \subseteq B^{\mathrm{cl}}$ together with the relations we already obtained for $(A \cup B)^{\mathrm{cl}}$ and $(A \cup B)^\circ$. This completes the fourth part. For the fifth we argue dually: $A \cap B \subseteq A, B$ shows $(A \cap B)^\circ \subseteq A^\circ \cap B^\circ$. However $A^\circ \subseteq A$ and $B^\circ \subseteq B$ gives $A^\circ \cap B^\circ \subseteq A \cap B$ and hence $A^\circ \cap B^\circ$ is an open subset inside $A \cap B$, the largest with this property is $(A \cap B)^\circ$. Hence they coincide. Next $A \cap B \subseteq A, B$ gives $(A \cap B)^{\mathrm{cl}} \subseteq A^{\mathrm{cl}} \cap B^{\mathrm{cl}}$ at once, showing the fifth part. For the last part, we first notice that the definition of a boundary point of A is symmetric in A and $M \setminus A$. Thus $\partial(M \setminus A) = \partial A$ follows immediately. Now suppose $p \in M \setminus A^\circ$. This is equivalent to $p \notin A^\circ$ and thus equivalent to $A \notin \mathfrak{U}(p)$. But A is not a neighbourhood of p iff for all neighbourhoods $U \in \mathfrak{U}(p)$ of p we have $U \cap (M \setminus A) \neq \emptyset$. This means that $p \in (M \setminus A)^{\mathrm{cl}}$ showing the equality $(M \setminus A)^{\mathrm{cl}} = M \setminus A^\circ$. Using this we get $M \setminus (M \setminus A)^\circ = (M \setminus (M \setminus A))^{\mathrm{cl}} = A^{\mathrm{cl}}$ and thus $M \setminus A^{\mathrm{cl}} = (M \setminus A)^\circ$. □

In Exercise 2.7.9 one can find examples of subsets where all the above inclusions are shown to be proper: thus they cannot be improved in general.

2.4 Continuous Maps

We come now to the central definition of continuity of maps. Motivated by the considerations in Proposition 2.1.8 one states the following definition:

Definition 2.4.1 (*Continuity*) Let $f: (M, \mathcal{M}) \longrightarrow (N, \mathcal{N})$ be a map between topological spaces.

(i) The map f is called continuous at $p \in M$ if for every neighbourhood $U \in \mathfrak{U}(f(p))$ of $f(p)$ also $f^{-1}(U)$ is a neighbourhood of p.
(ii) The map f is called continuous if the preimage of every open subset of N is open in M.
(iii) The set of continuous maps will be denoted by

$$\mathscr{C}(M, N) = \{f: M \longrightarrow N \mid f \text{ is continuous}\}, \qquad (2.4.1)$$

and we set $\mathscr{C}(M) = \mathscr{C}(M, \mathbb{C})$ for the complex-valued continuous functions on M.

We know by Proposition 2.1.8 that this reproduces the $\epsilon\delta$-continuity for metric spaces. Moreover, also in this more general situation the two notions are consistent in the following sense:

Proposition 2.4.2 *Let $f: (M, \mathcal{M}) \longrightarrow (N, \mathcal{N})$ be a map between topological spaces. Then the following statements are equivalent:*

(i) The map f is continuous at every point.
(ii) The map f is continuous.
(iii) The subset $f^{-1}(A) \subseteq M$ is closed for every closed $A \subseteq N$.
(iv) The subset $f^{-1}(\mathcal{O})$ is open for every $\mathcal{O} \in \mathcal{S}$ in a subbasis $\mathcal{S}$ of N.

Proof The equivalence of (ii) and (iii) is clear by taking complements. Assume (i), and let $\mathcal{O} \subseteq N$ be open and let $p \in f^{-1}(\mathcal{O})$. Then $f(p) \in \mathcal{O}$ shows $\mathcal{O} \in \mathfrak{U}(f(p))$ and thus $f^{-1}(\mathcal{O}) \in \mathfrak{U}(p)$. Since this holds for all $p \in f^{-1}(\mathcal{O})$, we have $f^{-1}(\mathcal{O})$ open. This gives (i) $\implies$ (ii). Conversely, suppose (ii), and let $U \in \mathfrak{U}(f(p))$. Then there is an open $\mathcal{O} \subseteq U$ with $f(p) \in \mathcal{O}$ and hence $p \in f^{-1}(\mathcal{O}) \in \mathfrak{U}(p)$, since $f^{-1}(\mathcal{O})$ is open by continuity. But then $f^{-1}(\mathcal{O}) \subseteq f^{-1}(U)$ shows $f^{-1}(U) \in \mathfrak{U}(p)$ giving (ii) $\implies$ (i) Finally, the compatibility of $\bigcap$ and $\bigcup$ with preimages shows the equivalence of (ii) and (iv). $\square$

In particular, the fourth part is often very convenient for checking continuity as we can use a rather small and easy subbasis instead of the typically huge and complicated topology.

To show the efficiency of this definition of continuity we first prove the following statements on compositions of maps:

Proposition 2.4.3 *Let $f: (M, \mathcal{M}) \longrightarrow (N, \mathcal{N})$ and $g: (N, \mathcal{N}) \longrightarrow (K, \mathcal{K})$ be maps between topological spaces.*

(i) If f is continuous at $p \in M$ and g is continuous at $f(p) \in N$ then $g \circ f$ is continuous at p.
(ii) If f and g are continuous then $g \circ f$ is continuous.

Proof Both statements rely on the simple fact that the preimage maps

$$f^{-1} : 2^N \longrightarrow 2^M \quad \text{and} \quad g^{-1} : 2^K \longrightarrow 2^N$$

satisfy

$$(g \circ f)^{-1} = f^{-1} \circ g^{-1}.$$

Then the preimages of neighbourhoods are mapped to neighbourhoods and the preimages of open subsets are mapped again to open subsets.   □

Remark 2.4.4 The very same argument is used in measure theory to show that the composition of measurable maps is again measurable.

While the preimages of open or closed subsets behave nicely under continuous maps, the images will show no particularly simple behaviour in general. The image of the continuous map

$$f : \mathbb{R} \ni x \ \mapsto \ \frac{1}{1 + x^2} \in \mathbb{R} \tag{2.4.2}$$

of the open (or closed) subset $\mathbb{R}$ is the half-open interval $(0, 1]$. Thus the following definitions provide additional features of maps:

Definition 2.4.5 (*Open and closed maps*) Let $f : (M, \mathcal{M}) \longrightarrow (N, \mathcal{N})$ be a map between topological spaces.

(i) The map f is called open if $f(\mathcal{O}) \subseteq N$ is open for all open $\mathcal{O} \subseteq M$.
(ii) The map f is called closed if $f(A) \subseteq N$ is closed for every closed subset $A \subseteq M$.

Suppose that a point $q \in N$ yields a closed subset $\{q\} \subseteq N$ then a constant map $f : M \ni p \mapsto q \in N$ is always closed since it simply maps every subset of M to a closed subset. It is also continuous but typically not open unless the single point $\{q\}$ is also an open subset of N. Consider the projection $\mathrm{pr}_1 : \mathbb{R}^2 \longrightarrow \mathbb{R}$ onto the first component. This is an open map but not a closed map, see Fig. 2.4. Thus the notions of continuous, open, and closed maps are rather independent, see also Exercise 2.7.18.

Finally, we introduce the notion of "isomorphism" between topological spaces. Even though isomorphism would be a conceptually more appropriate name, we stick to the traditional notion:

Definition 2.4.6 (*Homeomorphism*) Let $f : (M, \mathcal{M}) \longrightarrow (N, \mathcal{N})$ be a map between topological spaces.

(i) The map f is called a homeomorphism if f is bijective, continuous, and if f^{-1} is continuous.
(ii) If there is a homeomorphism $f : (M, \mathcal{M}) \longrightarrow (N, \mathcal{N})$ then the spaces $(M, \mathcal{M})$ and $(N, \mathcal{N})$ are called homeomorphic.

Fig. 2.4 The image of a
closed curve in $\mathbb{R}^2$ under the
continuous projection onto the
first coordinates may be open

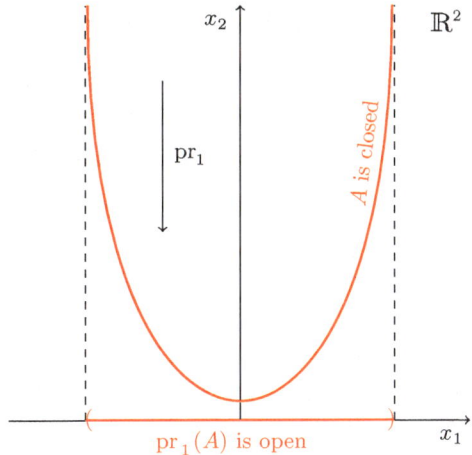

(iii) The map f is called an embedding if f is injective and if

$$f : (M, \mathcal{M}) \longrightarrow \left(f(M), \mathcal{N}\big|_{f(M)} \right) \tag{2.4.3}$$

is a homeomorphism.

In Exercise 2.7.15 we have an example that the inverse of a continuous bijection needs not to be continuous at all. Hence the requirement of the continuity of f^{-1} in the definition is *not* superfluous. Moreover, being homeomorphic is clearly an equivalence relation: the definition is symmetric in M and N and the composition of homeomorphisms is again a homeomorphism. Finally, we note that an injective continuous map is in general *not* an embedding, see again Exercise 2.7.15.

Example 2.4.7 Let $(M, \mathcal{M})$ be a topological space and let $N \subseteq M$ be a subset. Endow N with the subspace topology $\mathcal{N} = \mathcal{M}\big|_N$. Then the canonical inclusion map

$$\iota : N \longrightarrow M, \tag{2.4.4}$$

which identifies the points of N as (particular) points of M, is continuous and even an embedding. Indeed, for $\mathcal{O} \subseteq M$ open the preimage of $\mathcal{O}$ is $\iota^{-1}(\mathcal{O}) = N \cap \mathcal{O}$ which is open in $\mathcal{N}$ by the very definition of the subspace topology. Moreover, ι is clearly a bijection onto its image. Finally, the homeomorphism property is clear by the very definition. This example is the prototype of an embedding and motivates the name.

The next proposition gives some useful equivalent characterizations of homeomorphisms:

Proposition 2.4.8 *Let* $f : (M, \mathcal{M}) \longrightarrow (N, \mathcal{N})$ *be a map between topological spaces. Then the following statements are equivalent:*

(i) The map f is a homeomorphism.
(ii) The map f is continuous, bijective, and open.
(iii) The map f is continuous, bijective, and closed.
(iv) The map f is continuous and there exists a continuous map $g \colon N \longrightarrow M$ with $g \circ f = \mathsf{id}_M$ and $f \circ g = \mathsf{id}_N$.

Proof The equivalence of (i) and (iv) is clear by taking $g = f^{-1}$. Suppose (i) and let $\mathcal{O} \subseteq M$ be open. Since f^{-1} is continuous, $(f^{-1})^{-1}(\mathcal{O}) \subseteq N$ is open. But this is just $f(\mathcal{O})$, showing that f is open. Note that here we have two different meanings of "-1": we take the preimage of $\mathcal{O}$ under the inverse map. Analogously, one shows (i) $\implies$ (iii). Now assume (ii) and let $\mathcal{O} \subseteq M$ be open. Then $f(\mathcal{O}) \subseteq N$ is open as f is an open map. But again $f(\mathcal{O})$ is the preimage of $\mathcal{O}$ under the inverse map of f^{-1}, showing that f^{-1} is continuous. Hence (ii) $\implies$ (i) follows. Finally, (iii) $\implies$ (i) is again analogous. $\square$

Let us conclude this section with some more conceptual aspects. Recall that a *category* $\mathfrak{C}$ consists of a class of *objects* $\mathsf{Obj}(\mathfrak{C})$ and a set $\mathsf{Morph}(a, b)$ for any two objects $a, b \in \mathsf{Obj}(\mathfrak{C})$, the *morphisms* from a to b, such that one has a *composition*

$$\circ \colon \mathsf{Morph}(b, c) \times \mathsf{Morph}(a, b) \longrightarrow \mathsf{Morph}(a, c) \qquad (2.4.5)$$

and a *unit morphism* $\mathsf{id}_a \in \mathsf{Morph}(a, a)$ such that the composition of morphisms is associative whenever it is defined and id_a serves as unit for the composition whenever it can be composed. More background information on the theory of categories can e.g. be found in [24]. It is common and useful to depict the morphisms as arrows between the objects which can be composed whenever the tail and the head match. There are many examples of categories in mathematics: in general one can say that whenever one introduces a new type of structure on certain sets one should immediately ask for the structure preserving maps. Together this should yield a category.

Example 2.4.9 Without verifying the properties of a category, we just list some well-known examples:

(i) The category Set of sets with maps as arrows between them.
(ii) The category Group of groups with group homomorphisms as arrows between them.
(iii) The category Ring of unital rings with unital ring homomorphisms as arrows between them.
(iv) The category of complex vector spaces $\mathsf{Vect}_{\mathbb{C}}$ with linear maps as arrows between them.
(v) The trivial category $\mathbf{1}$ with one object 1 and one arrow id_1.

In a category $\mathfrak{C}$ one calls two objects $a, b \in \mathsf{Obj}(\mathfrak{C})$ *isomorphic* if there are morphisms $f \in \mathsf{Morph}(a, b)$ and $g \in \mathsf{Morph}(b, a)$ with $f \circ g = \mathsf{id}_b$ and $g \circ f = \mathsf{id}_a$. Clearly, in all the above examples this gives then the correct notion of isomorphisms.

The conclusion of this section can now be rephrased as follows: we have found a category of topological spaces with continuous maps as morphisms between them:

Proposition 2.4.10 *The topological spaces form a category* top *with respect to the continuous maps as morphisms between them. The isomorphisms in* top *are precisely the homeomorphisms.*

Proof The main point is that the composition of continuous maps is again continuous and that the identity map $\mathrm{id}_M : M \longrightarrow M$ is continuous, too. The associativity is always fulfilled for compositions of maps. Finally, the homeomorphisms are the isomorphisms thanks to Proposition 2.4.8, (iv). □

2.5 Connectedness

In this short section we discuss some further easy properties of topological spaces: connectedness and path-connectedness. The motivation of the definition of a connected topological space originates form the following observation:

Lemma 2.5.1 *Consider the closed interval* $M = [0, 1]$ *with its usual topology. Suppose we have two open subsets* $\mathcal{O}_1, \mathcal{O}_2 \subseteq [0, 1]$ *with* $\mathcal{O}_1 \cup \mathcal{O}_2 = [0, 1]$ *and* $\mathcal{O}_1 \cap \mathcal{O}_2 = \emptyset$. *Then necessarily* $\mathcal{O}_1$ *and* $\mathcal{O}_2$ *are just* $[0, 1]$ *and* $\emptyset$.

Proof Suppose we have two such open subsets $\mathcal{O}_1, \mathcal{O}_2$ in $[0, 1]$, both non-empty. Without restriction we find $x \in \mathcal{O}_1$ and $y \in \mathcal{O}_2$ such that $0 < x < y < 1$. Indeed, the openness of $\mathcal{O}_1$ and $\mathcal{O}_2$ allows to find more points than just the boundary points 0 and 1 inside $\mathcal{O}_1$ and $\mathcal{O}_2$. Now consider all those numbers $\xi \in [0, 1]$ with

$$[x, \xi] \subseteq \mathcal{O}_1$$

and define z to be their supremum. Since $[0, 1]$ is closed, $z \in [0, 1]$. Moreover, since $0 < x$ we get $0 < z$ and since $y < 1$ we also get $z < 1$, again by the openness of $\mathcal{O}_1$ and $\mathcal{O}_2$. Suppose $z \in \mathcal{O}_1$ then also the open interval $(z - \varepsilon, z + \varepsilon) \subseteq \mathcal{O}_1$ is in $\mathcal{O}_1$ for some small enough $\varepsilon > 0$ since $\mathcal{O}_1$ is open. But in this case $[x, z + \frac{\varepsilon}{2}] \subseteq \mathcal{O}_1$ contradicting the supremum property of z. Thus $z \in \mathcal{O}_2$ as $\mathcal{O}_1 \cup \mathcal{O}_2$ is the whole interval. But then again $(z - \varepsilon, z + \varepsilon) \subseteq \mathcal{O}_2$ by openness of $\mathcal{O}_2$ for some small $\varepsilon > 0$. Hence $z - \frac{\varepsilon}{2} \in \mathcal{O}_2$ can not be in $\mathcal{O}_1$, contradicting again the supremum property of z. This is the final contradiction yielding the proof. □

Definition 2.5.2 (*Connectedness*) Let $(M, \mathcal{M})$ be a topological space. Then M is called connected if there are no two open, disjoint subsets $\mathcal{O}_1, \mathcal{O}_2 \subseteq M$ with $\mathcal{O}_1 \cup \mathcal{O}_2 = M$ beside M and $\emptyset$. A subset $A \subseteq M$ is called connected if $(A, \mathcal{M}|_A)$ is connected.

Corollary 2.5.3 *The unit interval* $[0, 1]$ *is connected.*

With an analogous argument one shows that all other types of intervals in $\mathbb{R}$ like $[a, b], (a, b], [a, b),$ and (a, b) for $-\infty \leq a \leq b \leq \infty$ are connected, too. In fact, these are the only subsets of $\mathbb{R}$ which are connected:

Proposition 2.5.4 *Let $A \subseteq \mathbb{R}$ and $a, b \in A$. If A is connected, then $[a, b] \subseteq A$.*

Proof Suppose $z \in [a, b]$ does not belong to A. Then $(-\infty, z)$ and (z, ∞) are both open subsets of $\mathbb{R}$ and hence $\mathcal{O}_1 = A \cap (-\infty, z)$ as well as $\mathcal{O}_2 = A \cap (z, \infty)$ are open in the subspace topology of A. By assumption $\mathcal{O}_1 \cup \mathcal{O}_2 = A$, $\mathcal{O}_1 \cap \mathcal{O}_2 = \emptyset$, and $a \in \mathcal{O}_1$ while $b \in \mathcal{O}_2$ as z is different from a and b. This contradicts the connectedness of A. $\square$

Connectedness behaves well under continuous maps:

Proposition 2.5.5 *Let $f : (M, \mathcal{M}) \longrightarrow (N, \mathcal{N})$ be a continuous map between topological spaces. If M is connected then $f(M)$ is connected, too.*

Proof Suppose $f(M)$ is not connected and let $\mathcal{O}_1, \mathcal{O}_2 \subseteq f(M)$ be open and disjoint with $f(M) = \mathcal{O}_1 \cup \mathcal{O}_1$ but $\mathcal{O}_1, \mathcal{O}_2$ both be non-empty. Then we find $U_1, U_2 \subseteq N$ open with $\mathcal{O}_1 = f(M) \cap U_1$ and $\mathcal{O}_2 = f(M) \cap U_2$. Moreover, $f^{-1}(U_k) = f^{-1}(U_k \cap f(M)) = f^{-1}(\mathcal{O}_k)$ for $k = 1, 2$ shows that $f^{-1}(\mathcal{O}_1)$ and $f^{-1}(\mathcal{O}_2)$ are open in M, both non-empty, and still disjoint with $f^{-1}(\mathcal{O}_1) \cup f^{-1}(\mathcal{O}_2) = M$. But this contradicts the connectedness of M. $\square$

This simple fact together with the result of Proposition 2.5.4 can be seen as the topological "reason" for the intermediate value theorem in calculus: The continuous image of an interval is again an interval.

Corollary 2.5.6 (Intermediate value theorem) *Let $f : (M, \mathcal{M}) \longrightarrow \mathbb{R}$ be a continuous function on a connected topological space. If $a, b \in f(M)$ then also $[a, b] \subseteq f(M)$.*

Connectedness can also be understood by the idea that we can join any two points by a continuous path. This motivates the following definition:

Definition 2.5.7 (*Path-Connectedness*) Let $(M, \mathcal{M})$ be a topological space.

   (i) A path in M is a continuous map $\gamma : [0, 1] \longrightarrow M$.
   (ii) The space M is called path-connected if for any $p, q \in M$ one finds a path γ with

$$\gamma(0) = p \quad \text{and} \quad \gamma(1) = q. \tag{2.5.1}$$

Since we can reparametrize the "time" variable t of a path, there is no real restriction in requiring the domain of definition to be $[0, 1]$ instead of $[a, b] \subseteq \mathbb{R}$ for some $a < b$. We have now the following statement:

Proposition 2.5.8 *Let $(M, \mathcal{M})$ be a path-connected topological space. Then M is connected, too.*

Proof Suppose M is not connected and let $\mathcal{O}_1, \mathcal{O}_2 \subseteq M$ be open, disjoint, $M = \mathcal{O}_1 \cup \mathcal{O}_2$ with both $\mathcal{O}_1, \mathcal{O}_2$ being non-empty. Then let $p \in \mathcal{O}_1$ and $q \in \mathcal{O}_2$ and join them by a continuous path $\gamma : [0, 1] \longrightarrow M$, i.e. $\gamma(0) = p$ and $\gamma(1) = q$. Then $\gamma^{-1}(\mathcal{O}_1)$, $\gamma^{-1}(\mathcal{O}_2)$ are open, disjoint, both non-empty as $0 \in \gamma^{-1}(\mathcal{O}_1)$, $1 \in \gamma^{-1}(\mathcal{O}_2)$ and $[0, 1] = \gamma^{-1}(\mathcal{O}_1) \cup \gamma^{-1}(\mathcal{O}_2)$ since $\mathcal{O}_1 \cup \mathcal{O}_2 = M$. This contradicts the connectedness of $[0, 1]$. $\square$

In general, the reverse implication is not true: there are connected spaces which are not path-connected, see Exercise 2.7.23. Moreover, since the compositions of continuous maps are continuous, it is trivial to see that the image of a path-connected topological space under a continuous map is again path-connected.

If $(M, \mathcal{M})$ is not (path-)connected we can still ask for the largest subset containing a given point $p \in M$ which is (path-)connected. These subsets are characterized in the following Proposition:

Proposition 2.5.9 *Let $(M, \mathcal{M})$ be a topological space.*

(i) *If $\{C_i\}_{i \in I}$ is a family of (path-)connected subsets of M such that $\bigcap_{i \in I} C_i \neq \emptyset$ then $\bigcup_{i \in I} C_i$ is again (path-)connected.*

(ii) *If $A \subseteq B \subseteq A^{\mathrm{cl}} \subseteq M$ and A is a connected subset then B is connected as well. In particular, A^{cl} is connected.*

(iii) *The union $\mathfrak{C}(p)$ of all connected subsets of M which contain p is connected and closed.*

(iv) *The union $\Pi(p)$ of all path-connected subsets of M which contain p is path-connected and*

$$\Pi(p) \subseteq \mathfrak{C}(p). \tag{2.5.2}$$

Proof Let $C = \bigcup_{i \in I} C_i$. Moreover, let $\mathcal{O}_1, \mathcal{O}_2 \subseteq M$ be open subsets with $(\mathcal{O}_1 \cap C) \cap (\mathcal{O}_2 \cap C) = \emptyset$ and $C \subseteq \mathcal{O}_1 \cup \mathcal{O}_2$. Since $C_i \subseteq C$ we have $\mathcal{O}_1 \cup \mathcal{O}_2 \supseteq C_i$ for all $i \in I$. For a fixed $i_0 \in I$ we have $\mathcal{O}_1 \cap C_{i_0}$ or $\mathcal{O}_2 \cap C_{i_0}$ empty by the connectedness of C_{i_0}. Without restriction, we can assume $\mathcal{O}_2 \cap C_{i_0} = \emptyset$ and thus $C_{i_0} \subseteq \mathcal{O}_1 \cap C_{i_0}$. But then the non-empty set $\bigcap_{j \in I} C_j$ is also contained in $\mathcal{O}_1 \cap C_{i_0}$ and hence $C_j \cap \mathcal{O}_1 \neq \emptyset$ for all $j \in I$. By the connectedness of all the C_j we conclude that $C_j \subseteq \mathcal{O}_1 \cap C_j$ for all $j \in I$ and thus $C \subseteq \mathcal{O}_1 \cap C$ proving that C is connected. The path-connected case is easier by joining two points $p, q \in C$ with $p \in C_i$ and $q \in C_j$ first to a common point $x \in \bigcap_{i \in I} C_i$ and reparametrizing the joined paths afterwards. Note that this gives indeed a continuous path again, see also Exercise 2.7.21. For the second part, let $A \subseteq B \subseteq A^{\mathrm{cl}}$ be given, with A being connected. Suppose B is not connected. Then we find two open subsets $\mathcal{O}_1, \mathcal{O}_2 \subseteq M$ with $B \subseteq \mathcal{O}_1 \cup \mathcal{O}_2$, $(B \cap \mathcal{O}_1) \cap (B \cap \mathcal{O}_2) = \emptyset$ and $B \cap \mathcal{O}_1 \neq \emptyset \neq B \cap \mathcal{O}_2$. Then also $A \subseteq \mathcal{O}_1 \cup \mathcal{O}_2$ and $(A \cap \mathcal{O}_1) \cap (A \cap \mathcal{O}_2) = \emptyset$. Since $B \subseteq A^{\mathrm{cl}}$ we have for every $p \in B$ and every open subset $\mathcal{O}$ with $p \in \mathcal{O}$, a non-trivial intersection $A \cap \mathcal{O} \neq \emptyset$. Choosing $b_1 \in B \cap \mathcal{O}_1$ and $b_2 \in B \cap \mathcal{O}_2$ shows $A \cap \mathcal{O}_1 \neq \emptyset \neq A \cap \mathcal{O}_2$, a contradiction to the connectedness of A. The second part implies that for every connected $C \subseteq M$ also C^{cl} is connected. Thus we conclude that $\mathfrak{C}(p) = \mathfrak{C}(p)^{\mathrm{cl}}$ is again connected. For the last part we consider all path-connected subsets containing p. Their intersection is non-empty as it still contains p. Hence we can apply the first part. Since a path-connected subset is also connected by Proposition 2.5.8, the conclusion follows. $\square$

In general, the closure of a path-connected subset is no longer path-connected but only connected, see Exercise 2.7.23. The subsets $\mathfrak{C}(p)$ and $\Pi(p)$ deserve particular attention:

Definition 2.5.10 (*Connected components*) Let $(M, \mathcal{M})$ be a topological space and $p \in M$.

(i) The subset $\mathfrak{C}(p)$ is called the connected component of p.
(ii) The subset $\Pi(p)$ is called the path-connected component of p.

It is now fairly easy to see that $q \in \mathfrak{C}(p)$ holds iff $p \in \mathfrak{C}(q)$. Moreover, if $q \in \mathfrak{C}(p)$ and $x \in \mathfrak{C}(q)$ then $x \in \mathfrak{C}(p)$. Thus we get an equivalence relation of *being in the same connected component* of M, see Exercise 2.7.22. The same holds for the path-connected components.

The connectedness is a global feature of a topological space, however, many properties relying on connectedness can also hold true if the connectedness is satisfied only locally. This yields the definition of locally (path-)connected spaces. One version to formulate the other extreme case is the notion of totally disconnected spaces, see also e.g. [32, Part I, Sect. 4] for many further notions of (dis-)connectedness.

Definition 2.5.11 (*Local connectedness and total disconnectedness*) Let $(M, \mathcal{M})$ be a topological space.

(i) If for every point $p \in M$ every neighbourhood $U \in \mathfrak{U}(p)$ contains a (path-) connected neighbourhood of p then $(M, \mathcal{M})$ is called locally (path-) connected.
(ii) If $\mathfrak{C}(p) = \{p\}$ for all $p \in M$ then $(M, \mathcal{M})$ is called totally disconnected.

Some illustrating examples of these extreme cases are discussed in Exercise 2.7.23 and Exercise 2.7.24.

Proposition 2.5.12 *Let $(M, \mathcal{M})$ be a topological space.*

(i)   *If M is locally connected then the connected component $\mathfrak{C}(p)$ of $p \in M$ is open.*
(ii) *The space M is locally connected iff the connected open subsets form a basis of the topology.*
(iii) *If M is locally path-connected then for all $p \in M$ we have*

$$\mathfrak{C}(p) = \Pi(p). \tag{2.5.3}$$

(iv) *Suppose M is locally path-connected. Then M is connected iff M is path-connected.*

Proof Let $p, q \in M$ with $q \in \mathfrak{C}(p)$ be given. Then for a connected neighbourhood U of q we have $U \cap \mathfrak{C}(p) \neq \emptyset$ since q belongs to this intersection. By Proposition 2.5.9, (i), we have that $U \cup \mathfrak{C}(p)$ is still connected, hence $U \subseteq \mathfrak{C}(p)$ follows since $\mathfrak{C}(p)$ is the largest connected subset containing p. Thus $\mathfrak{C}(p)$ is a neighbourhood of q and thus open by Proposition 2.3.2, (i), proving the first part. If M is locally connected, then the open connected neighbourhoods of a point form a basis of neighbourhoods of that point. Hence they also form a basis of the topology. The converse is true by the same line of argument. Now let M be locally path-connected and hence locally connected. An analogous argument as for the first part shows that $\Pi(p)$ is open for every $p \in M$. From Proposition 2.5.9, (iv), we have $\Pi(p) \subseteq \mathfrak{C}(p)$. Suppose

$q \in \mathfrak{C}(p) \setminus \Pi(p)$. Then $\Pi(q) \subseteq \mathfrak{C}(p) \setminus \Pi(p)$ since if $\Pi(q) \cap \Pi(p) \neq \emptyset$ we would have already $\Pi(q) = \Pi(p)$. This shows

$$\mathfrak{C}(p) \setminus \Pi(p) = \bigcup_{q \in \mathfrak{C}(p) \setminus \Pi(p)} \Pi(q),$$

and hence $\mathfrak{C}(p) \setminus \Pi(p)$ is open. This gives a non-trivial decomposition of $\mathfrak{C}(p)$ into the disjoint open subsets $\Pi(p)$ and $\mathfrak{C}(p) \setminus \Pi(p)$. Since $\mathfrak{C}(p)$ is connected, one of them has to be empty. Since $p \in \Pi(p)$, we have $\mathfrak{C}(p) \setminus \Pi(p) = \emptyset$, which is the third part. The fourth is then a trivial consequence. $\square$

Example 2.5.13 An open subset $\mathcal{O} \subseteq \mathbb{R}^n$ is locally path-connected since clearly every open ball $B_r(x) \subseteq \mathcal{O}$ is path-connected: the straight line $[0, 1] \ni t \mapsto (1 - t) x + ty \in B_r(x)$ for $y \in B_r(x)$ connects x and y. Thus the notions of connectedness and path-connectedness for open subsets in $\mathbb{R}^n$ coincide.

2.6 Separation Properties

In a totally disconnected space it follows that all points constitute closed subsets $\{p\} \subseteq M$. Of course, there are many other topological spaces with this property without being totally disconnected as e.g. $\mathbb{R}$ with its standard topology. The following separation properties or "axioms" collect such features of how points in a topological space can be separated from each other.

Definition 2.6.1 (*Separation properties*) Let $(M, \mathcal{M})$ be a topological space.

  (i)   The space M is called a T_0-space if for each two different points $p \neq q$ in M we find an open subset which contains only one of them.
 (ii)   The space M is called a T_1-space if for each two different points $p \neq q$ in M we find open subsets $\mathcal{O}_1$ and $\mathcal{O}_2$ with $p \in \mathcal{O}_1$ and $q \in \mathcal{O}_2$ but $p \notin \mathcal{O}_2$ and $q \notin \mathcal{O}_1$.
(iii)   The space M is called a T_2-space or a Hausdorff space if for each two different points $p \neq q$ in M we find disjoint open subsets $\mathcal{O}_1$ and $\mathcal{O}_2$ with $p \in \mathcal{O}_1$ and $q \in \mathcal{O}_2$.
 (iv)   The space M is called a T_3-space if for every closed subset $A \subseteq M$ and every $p \in M \setminus A$ there are disjoint open subsets $\mathcal{O}_1$ and $\mathcal{O}_2$ with $A \subseteq \mathcal{O}_1$ and $p \in \mathcal{O}_2$.
  (v)   The space M is called a T_4-space if for two disjoint closed subsets $A_1, A_2 \subseteq M$ there are disjoint open subsets $\mathcal{O}_1$ and $\mathcal{O}_2$ with $A_1 \subseteq \mathcal{O}_1$ and $A_2 \subseteq \mathcal{O}_2$.

These are only the most common and important separation properties, many more like $T_{2\frac{1}{2}}$ etc. can be found in e.g. [27, Sect. 6A] or [32, Part I, Sect. 2]. The heuristic meaning of these properties can easily be visualized, see Fig. 2.5.

Example 2.6.2 Consider $M = \mathbb{R}$ with the following non-standard topology: let $\mathcal{O} \subseteq \mathbb{R}$ be open if $\mathcal{O} = (-\infty, a)$ for some $a \in \mathbb{R}$ or $\mathcal{O} = \mathbb{R}, \emptyset$. This indeed defines

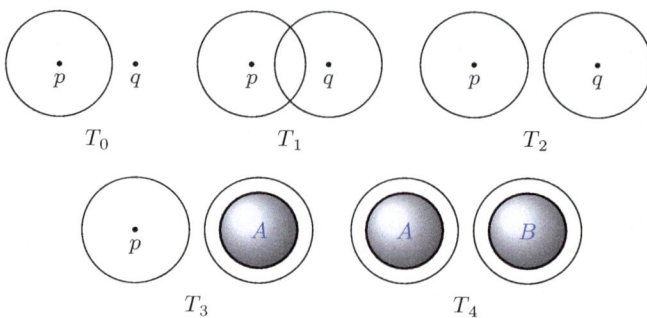

Fig. 2.5 The separation properties

a topology on $\mathbb{R}$. For two points $x, y \in \mathbb{R}$ with $x \neq y$ we have without restriction $x < y$. Then there is an open neighbourhood $\mathcal{O} = (-\infty, x + \varepsilon)$ of x with $y \notin \mathcal{O}$ by taking $\varepsilon > 0$ small enough. However, every open subset containing y also contains x. This shows that T_0 in general does not imply T_1. Of course, T_1 implies T_0.

Example 2.6.3 For the cofinite topology, the space $\mathbb{R}$ is a T_1-space: indeed for $x \neq y$, the subsets $\mathcal{O}_1 = \mathbb{R} \setminus \{y\}$ and $\mathcal{O}_2 = \mathbb{R} \setminus \{x\}$ will do the job. However, it is not a T_2-space as all non-empty open subsets overlap non-trivially. Hence T_1 does not imply T_2 but of course T_2 implies T_1. Since the only closed subsets are the finite ones, it is also easy to see that the cofinite topology does neither fulfill T_3 nor T_4.

Since single points need not to be closed, neither T_3 nor T_4 implies T_1 or T_2. Many examples and counterexamples can be found in [32].
    We collect now some useful reformulations and simple implications between combinations of the separation axioms.

Proposition 2.6.4 *A topological space $(M, \mathcal{M})$ is a T_1-space iff every point $p \in M$ gives a closed subset $\{p\} \subseteq M$.*

Proof The T_1-property means that for $p \in M$ the complement of p is a neighbourhood of any point in the complement. This is equivalent to $M \setminus \{p\}$ is open and thus $\{p\}$ is closed.   □

Proposition 2.6.5 *A topological space $(M, \mathcal{M})$ is a T_3-space iff for every $p \in M$ and every open $\mathcal{O} \in \mathfrak{U}(p)$ one finds an open $U \in \mathfrak{U}(p)$ with*

$$p \in U \subseteq U^{\mathrm{cl}} \subseteq \mathcal{O}. \tag{2.6.1}$$

This means that there is a neighbourhood basis of closed subsets for each point $p \in M$.

Proof Let $\mathcal{O} \in \mathfrak{U}(p)$ be open then $M \setminus \mathcal{O}$ is closed and $p \notin M \setminus \mathcal{O}$. Hence there are open subsets $V, U \subseteq M$ with $V \cap U = \emptyset$ and $p \in U$ as well as $M \setminus \mathcal{O} \subseteq V$, see Fig. 2.6. But then $M \setminus V$ is closed and

Fig. 2.6 The separating open
subsets V and U in a T_3-space

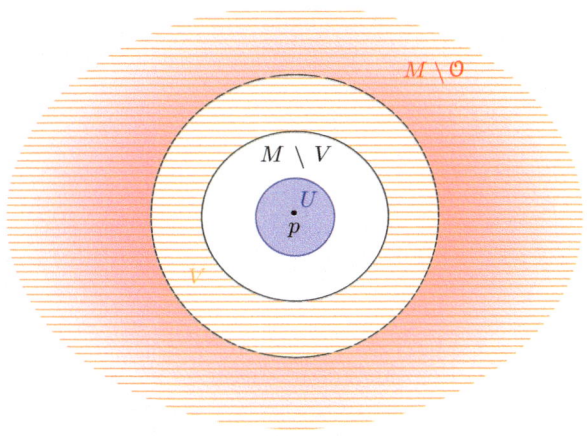

$$U \subseteq M \setminus V \subseteq \mathcal{O}.$$

Since U^{cl} is the smallest closed subset containing U we have $U^{\mathrm{cl}} \subseteq M \setminus V \subseteq \mathcal{O}$ and hence we have found U with (2.6.1). Conversely, assume (2.6.1) and let $A \subseteq M$ be closed and $p \notin A$. Then $p \in M \setminus A$ and $M \setminus A$ is open. Hence we can choose an open $U \subseteq U^{\mathrm{cl}} \subseteq M \setminus A$ with $p \in U$. Clearly $U^{\mathrm{cl}} \cap A = \emptyset$ and thus $A \subseteq M \setminus U^{\mathrm{cl}} = \mathcal{O}$. Then U and $\mathcal{O}$ will separate $\{p\}$ and A as required for T_3. $\square$

With an analogous argument one shows that the T_4 property can be formulated equivalently as follows:

Proposition 2.6.6 *A topological space $(M, \mathcal{M})$ is a T_4-space iff for every closed subset $A \subseteq M$ and every open $\mathcal{O} \subseteq M$ with $A \subseteq \mathcal{O}$ we have an open $U \subseteq M$ with*

$$A \subseteq U \subseteq U^{\mathrm{cl}} \subseteq \mathcal{O}. \tag{2.6.2}$$

Since the axioms T_3 and T_4 are quite unrelated to T_1 and T_2 it seems reasonable to require both: separation of points and separation of closed subsets. This motivates the following definition:

Definition 2.6.7 *(Regular and normal spaces)* Let $(M, \mathcal{M})$ be a topological space.

 (i) The space $(M, \mathcal{M})$ is called regular if it is T_1 and T_3.
(ii) The space $(M, \mathcal{M})$ is called normal if it is T_1 and T_4.

Proposition 2.6.8 *A regular space is Hausdorff and a normal space is regular.*

Proof Since by T_1 all points $\{p\} \subseteq M$ are closed, T_3 separates a point p from the closed subset $\{q\}$ for $p \neq q$ by disjoint open subsets. Thus T_2 follows. Again by T_1 a point is closed and hence T_4 implies T_3. $\square$

Thanks to this proposition it will be mainly the combination of T_2 with some of the remaining separation properties which will be of most importance. The normal spaces will enjoy several other nice properties, similar to metric spaces. Indeed, metric spaces are normal:

Proposition 2.6.9 *A metric space is normal and hence T_1, T_2, T_3, T_4.*

Proof Obviously, a single point $\{p\}$ is closed as $M \setminus \{p\} = \bigcup_{q \in M \setminus \{p\}} B_{r_q}(q)$ with $0 < r_q < d(q, p)$ is open. Now let $A, B \subseteq M$ be closed with $A \cap B = \emptyset$. For $p \in A$ we have a radius $r_p > 0$ with $B_{r_p}(p) \cap B = \emptyset$ since $p \in M \setminus B$ and $M \setminus B$ is open. Analogously, for $q \in B$ we find $r_q > 0$ with $B_{r_q}(q) \cap A = \emptyset$. Define the open subsets

$$U = \bigcup_{p \in A} B_{r_p/2}(p) \quad \text{and} \quad V = \bigcup_{q \in A} B_{r_q/2}(q).$$

Then $A \subseteq U$ and $B \subseteq V$ is clear. Moreover, since $r_p < d(p, q)$ for all $q \in B$ and $r_q < d(q, p)$ for all $p \in A$ we see that $B_{r_p/2}(p) \cap B_{r_q/2}(q) = \emptyset$, by the triangle inequality, see also Fig. 2.7. But then $U \cap V = \emptyset$ follows which gives T_4. $\square$

We will come back to the separation axioms at several instances. In particular, the existence of sufficiently non-trivial continuous functions relies heavily on the separation properties. As a last application of the Hausdorff property we mention the following two statements which turn out to be very useful at many places:

Proposition 2.6.10 *Let $f, g \colon (M, \mathcal{M}) \longrightarrow (N, \mathcal{N})$ be continuous maps between topological spaces and assume that $(N, \mathcal{N})$ is Hausdorff.*

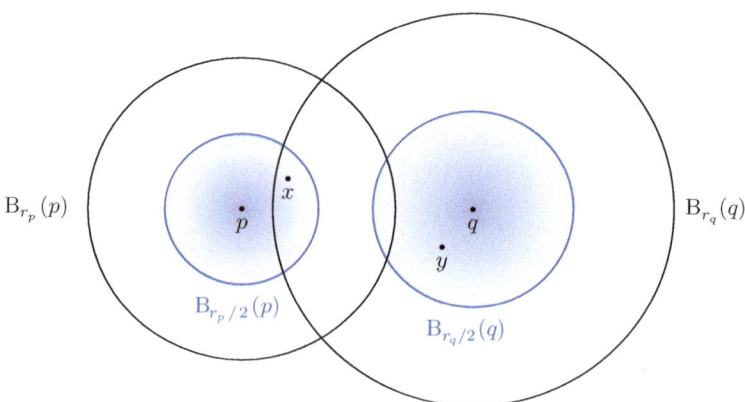

Fig. 2.7 A metric space is T_4

(i) The coincidence set $\{p \in M \mid f(p) = g(p)\} \subseteq M$ is closed.
(ii) If $U \subseteq M$ is dense then $f\big|_U = g\big|_U$ implies $f = g$.

Proof Let $q \in M$ be a point with $f(q) \neq g(q)$. Then the Hausdorff property implies that we find open subsets $\mathcal{O}_1, \mathcal{O}_2 \subseteq N$ with $f(q) \in \mathcal{O}_1$ and $g(q) \in \mathcal{O}_2$ but $\mathcal{O}_1 \cap \mathcal{O}_2 = \emptyset$. By continuity, $f^{-1}(\mathcal{O}_1)$ and $g^{-1}(\mathcal{O}_2)$ are open and $q \in f^{-1}(\mathcal{O}_1) \cap g^{-1}(\mathcal{O}_2)$. If q' is another point in this intersection then $f(q') \in \mathcal{O}_1$ and $g(q') \in \mathcal{O}_2$ yielding $f(q') \neq g(q')$ as $\mathcal{O}_1 \cap \mathcal{O}_2 = \emptyset$. This shows the first part by taking complements. The second part is now easy as the set $U \subseteq M$ is in the *closed* coincidence set and thus also U^{cl} is in the closed coincidence set. But $U^{\mathrm{cl}} = M$ is already everything. $\square$

This feature and many other pleasant properties of Hausdorff spaces justify to consider them with special care:

Definition 2.6.11 (*Hausdorff spaces*) The subcategory of top consisting of all Hausdorff topological spaces will be denoted by Top.

As the final remark we note that most of the separation properties behave nicely when passing to subspaces:

Proposition 2.6.12 *Let $(M, \mathcal{M})$ be a T_k-space with $k \in \{0, 1, 2, 3\}$ and let $N \subseteq M$ be endowed with the subspace topology. Then $(N, \mathcal{M}\big|_N)$ is T_k, too.*

Proof The proof is similar for all cases, we illustrate it for T_3: thus let $p \in N$ and $A \subseteq N$ be closed with $p \notin A$. We know that there is a closed $B \subseteq M$ with $A = N \cap B$ in this case, see Exercise 2.7.3, (ii). Since $p \in N$ we conclude that $p \notin B$, too. Hence we can separate $\{p\}$ and $B \in M$ by $U_1, U_2 \in \mathcal{M}$, i.e. $p \in U_1$, $B \subseteq U_2$, and $U_1 \cap U_2 = \emptyset$. But then $\mathcal{O}_1 = N \cap U_1$, $\mathcal{O}_2 = N \cap U_2$ are open in N and separate $\{p\}$ from A as wanted. $\square$

2.7 Exercises

Exercise 2.7.1 (Cartesian product of metric spaces) Let (M_1, d_1) and (M_2, d_2) be metric spaces. Consider their Cartesian product $M = M_1 \times M_2$ with

$$d((x_1, x_2), (y_1, y_2)) = d_1(x_1, y_1) + d_2(x_2, y_2), \tag{2.7.1}$$

$$d'((x_1, x_2), (y_1, y_2)) = \sqrt{d_1(x_1, y_1)^2 + d_2(x_2, y_2)^2}, \tag{2.7.2}$$

and

$$d''((x_1, x_2), (y_1, y_2)) = \max\{d_1(x_1, y_1), d_2(x_2, y_2)\}, \tag{2.7.3}$$

where $(x_1, x_2), (y_1, y_2) \in M_1 \times M_2$.

(i) Show that d, d', and d'' yield metrics on M.
(ii) Show that the open subsets with respect to all these three metrics d, d', and d'' coincide.
(iii) Show that the canonical projections

$$M_1 \xleftarrow{\text{pr}_1} M_1 \times M_2 \xrightarrow{\text{pr}_2} M_2 \tag{2.7.4}$$

are continuous.
(iv) Generalize these results to finite Cartesian products with more than two factors.

For a countable Cartesian product one can still construct a metric: let (M_n, d_n) be metric spaces for $n \in \mathbb{N}$. Then the Cartesian product $M = \prod_{n=1}^\infty M_n$ is the space of all sequences $x = (x_n)_{n\in\mathbb{N}}$, where the n-th term x_n is in M_n. One defines

$$d(x, y) = \sum_{n=1}^\infty \frac{1}{2^n} \frac{d_n(x_n, y_n)}{1 + d_n(x_n, y_n)}. \tag{2.7.5}$$

(v) Show that d is a well-defined metric on M and verify that the projection pr_n onto the n-th component is continuous for all $n \in \mathbb{N}$.
Hint: Use Example 2.1.2, (vi).

Exercise 2.7.2 (Formal power series) Consider the real formal power series $\mathbb{R}[[\lambda]]$ in a formal parameter λ. Let $o: \mathbb{R}[[\lambda]] \longrightarrow \mathbb{N}_0 \cup \{+\infty\}$ be the *order* of the power series as in Example 2.1.2, (iii), with the corresponding metric $d(a, b) = 2^{-o(a-b)}$ for $a, b \in \mathbb{R}[[\lambda]]$ where as usual we set $2^{-\infty} = 0$.

(i) Show that d is a metric for $\mathbb{R}[[\lambda]]$ satisfying the stronger version of the triangle inequality

$$d(a, b) \leq \max\{d(a, c), d(c, b)\} \tag{2.7.6}$$

for all $a, b, c \in \mathbb{R}[[\lambda]]$. A metric with this additional property is also called an *ultrametric*.
(ii) Endow $\mathbb{R}[[\lambda]] \times \mathbb{R}[[\lambda]]$ with one of the (equivalent) product metrics from Exercise 2.7.1 and show that the addition of formal power series as well as the multiplication defined by the Cauchy product

$$ab = \left(\sum_{n=0}^\infty \lambda^n a_n\right)\left(\sum_{m=0}^\infty \lambda^m b_m\right) = \sum_{k=0}^\infty \lambda^k \sum_{n+m=k} a_n b_m \tag{2.7.7}$$

are continuous.

(iii) Rephrase the condition for a Cauchy sequence in terms of the order and show that $\mathbb{R}[[\lambda]]$ is complete.

(iv) Show that the subspace topology of $\mathbb{R}$ induced from $\mathbb{R} \subseteq \mathbb{R}[[\lambda]]$ is the discrete topology. Show also that the topology of $\mathbb{R}[[\lambda]]$ is not the discrete one.

(v) Show that the polynomials $\mathbb{R}[\lambda] \subseteq \mathbb{R}[[\lambda]]$ are dense. More precisely, show that

$$\lim_{N \longrightarrow \infty} \sum_{n=0}^{N} \lambda^n a_n = \sum_{n=0}^{\infty} \lambda^n a_n \tag{2.7.8}$$

for every formal power series.

Exercise 2.7.3 (Subspace topology) Let $(M, \mathcal{M})$ be a topological space and let $N \subseteq M$ be a subset.

(i) Show that N is open in M iff every open subset $U \subseteq N$ with respect to the subspace topology $\mathcal{M}\big|_N$ is also open in M.

(ii) Show that $B \subseteq N$ is closed with respect to the subspace topology $\mathcal{M}\big|_N$ iff there is a closed subset $A \subseteq M$ with $B = A \cap N$.

(iii) Formulate and prove an analogous statement to (i) for closed subsets.

Exercise 2.7.4 (Neighbourhoods determine the topology) Let M be a set. For every point $p \in M$ consider a non-empty system of subsets $\mathfrak{U}(p)$ of M, such that the properties (ii), (iii), (iv), and (v) of Proposition 2.3.2 are satisfied.

(i) Define $\mathcal{O} \subseteq M$ to be open, if $\mathcal{O} \in \mathfrak{U}(p)$ for all $p \in \mathcal{O}$. Show that this defines a topology $\mathcal{M}$ on M.

(ii) Determine the neighbourhoods $\tilde{\mathfrak{U}}(p)$ of $p \in M$ for this topology $\mathcal{M}$ and show $\tilde{\mathfrak{U}}(p) = \mathfrak{U}(p)$ for all points $p \in M$.

This shows that the characterization of topological spaces via neighbourhood systems is equivalent to the characterization via topologies.

Exercise 2.7.5 (Finer and coarser topologies) Consider $\mathbb{R}$ with the discrete, the indiscrete, the cofinite, the topology from Example 2.6.2, and the standard (metric) topology. Order them with respect to being finer.

Exercise 2.7.6 (Rational balls) Consider $\mathbb{R}^n$ with its usual Euclidean metric and the corresponding topology. Show that every open subset can be obtained as union of open balls of the form $B_r(p)$ with $r \in \mathbb{Q}^+$ and $p \in \mathbb{Q}^n$.

Exercise 2.7.7 (Countability and subspaces) Let $(M, \mathcal{M})$ be a topological space and let $N \subseteq M$ be a subset being endowed with the subspace topology.

(i) Show that if $(M, \mathcal{M})$ is first countable at every point of N, then $(N, \mathcal{M}\big|_N)$ is first countable as well.

(ii) Show that if $(M, \mathcal{M})$ is second countable, then $(N, \mathcal{M}\big|_N)$ is second countable, too.

Exercise 2.7.8 (Closed and open subsets) Let $(M, \mathcal{M})$ be a topological space and let $A \subseteq M$ be a subset.

(i) Show that A is closed iff $A = A^{\mathrm{cl}}$ iff $\partial A \subseteq A$.
(ii) Show that A is open iff $A = A^{\circ}$.

Exercise 2.7.9 (Closures, open interiors, and boundaries) Find and describe examples of topological spaces $(M, \mathcal{M})$ and subsets $A, B \subseteq M$ for the following statements:

(i) The boundary of the boundary of a subset can but needs not to be empty.
(ii) Let $A \subseteq B$. Show that the following three situations are possible: a strict inclusion $\partial A \subseteq \partial B$, a strict inclusion $\partial B \subseteq \partial A$, a trivial intersection $\partial A \cap \partial B = \emptyset$ with both boundaries being non-empty.
(iii) The open interior of a union $A \cup B$ can be strictly larger than the union of the open interiors $A^{\circ} \cup B^{\circ}$.
(iv) The open interior of a boundary can be non-empty.
(v) The intersection of the boundaries $\partial A \cap \partial B$ can be strictly contained in the boundary of the intersection $A \cap B$.
(vi) The boundary of the intersection $A \cap B$ of two subsets can be strictly contained in the intersection of the boundaries of A and B.

Exercise 2.7.10 (Closure in the subspace topology) Let $(M, \mathcal{M})$ be a topological space and let $N \subseteq M$ be a subset endowed with the subspace topology $\mathcal{M}\big|_N$. Furthermore, let $A \subseteq N$ and denote the closure of A with respect to $\mathcal{M}\big|_N$ by $\overline{A}$. Show that $\overline{A} = A^{\mathrm{cl}} \cap N$.

Exercise 2.7.11 (Dense subsets) Let $(M, \mathcal{M})$ be a topological space and let $A \subseteq M$. Show that A is dense iff $A \cap \mathcal{O} \neq \emptyset$ for all non-empty open $\mathcal{O} \in \mathcal{M}$.

Exercise 2.7.12 (Nowhere dense subsets) Let $(M, \mathcal{M})$ be a topological space and let $A \subseteq M$ be a subset. Show that A is nowhere dense iff $(M \setminus A^{\mathrm{cl}})^{\mathrm{cl}} = M$ iff $M \setminus A^{\mathrm{cl}}$ is a dense open subset.

Exercise 2.7.13 (Density of $\mathbb{Q}^n$ in $\mathbb{R}^n$) Show that the set of rational points $\mathbb{Q}^n$ is dense in $\mathbb{R}^n$.

Exercise 2.7.14 (Topologies and continuous maps) Consider the sets $M_1 = \{1, 2\}$ und $M_2 = \{1, 2, 3\}$.

(i) Determine all topologies on M_1 and M_2.
(ii) Order the topologies on M_2 with respect to being finer. Can all of them be compared?
(iii) Determine all continuous maps $f : M_1 \longrightarrow M_2$ for all combinations for the topologies on M_1 and M_2, respectively.

Exercise 2.7.15 (Continuous maps and homeomorphisms) Consider the unit circle $\mathbb{S}^1 \subseteq \mathbb{C}$ and the half-open interval $[0, 2\pi)$, both endowed with their usual subspace topologies.

(i) Show that the map

$$f : [0, 2\pi) \ni t \mapsto e^{it} \in \mathbb{S}^1 \qquad (2.7.9)$$

is continuous.

(ii) Show that f is bijective.

(iii) Show that the inverse map $f^{-1} : \mathbb{S}^1 \longrightarrow [0, 2\pi)$ is not continuous.

Viewing f as a map from $[0, 2\pi)$ into $\mathbb{C}$ this gives also an example of a injective continuous map which is not an embedding.

Exercise 2.7.16 (Finer and coarser topologies and continuity) Let $f : (M, \mathcal{M}_1) \longrightarrow (N, \mathcal{N}_1)$ be a continuous map between topological spaces.

(i) Discuss whether or not f stays continuous if the topology $\mathcal{M}_1$ on the domain M is replaced by a finer (or coarser) topology $\mathcal{M}_2$, respectively.

(ii) Discuss whether or not f stays continuous if the topology $\mathcal{N}_1$ on the target N is replaced by a finer (or coarser) topology $\mathcal{N}_2$, respectively.

(iii) Show that the identity map id: $(M, \mathcal{M}_1) \longrightarrow (M, \mathcal{M}_2)$ is continuous iff $\mathcal{M}_1$ is finer than $\mathcal{M}_2$.

(iv) Show that the discrete topology on M is the unique topology $\mathcal{M}$ such that every map $f : (M, \mathcal{M}) \longrightarrow (N, \mathcal{N})$ into a topological space is continuous. Formulate and prove the analogous statement for the indiscrete topology.

Exercise 2.7.17 (Open map via a basis) Sometimes it is useful to characterize the openness of a map in terms of a basis of the topology: Show that a map $f : (M, \mathcal{M}) \longrightarrow (N, \mathcal{N})$ between topological spaces is open iff for a basis $\mathcal{B} \subseteq \mathcal{M}$ and all subsets $\mathcal{O} \in \mathcal{B}$ one has $f(\mathcal{O}) \in \mathcal{N}$.

Exercise 2.7.18 (Open, closed, and continuous maps)

(i) Find an example of a closed but discontinuous map $f : \mathbb{R} \longrightarrow \mathbb{R}$. Can you arrange it such that it is not open?

(ii) Find an example of an open but discontinuous map. The easiest way might be to use the discrete topology.

(iii) Consider finally the inclusion $\mathbb{R} \longrightarrow \mathbb{R}^2$ and the projection $\mathrm{pr}_1 : \mathbb{R}^2 \longrightarrow \mathbb{R}$ and discuss whether they are open, closed, or continuous.

Exercise 2.7.19 (The algebra $\mathscr{C}_b(M)$) Let $(M, \mathcal{M})$ be a topological space and consider the set $\mathscr{C}(M)$ of $\mathbb{C}$-valued continuous functions on M.

(i) Show that $\mathscr{C}(M)$ is a vector space with respect to the pointwise addition and multiplication with a scalar in $\mathbb{C}$. Show that the pointwise complex conjugation yields an involution on $\mathscr{C}(M)$.

(ii) Show that $\mathscr{C}(M)$ becomes a commutative associative algebra with unit with respect to the pointwise product.

(iii) Show that the maximum and the minimum of real-valued continuous functions as well as the absolute value of a continuous function are again continuous.

(iv) Consider now the $\mathbb{C}$-valued bounded continuous functions

$$\mathscr{C}_b(M) = \left\{ f \in \mathscr{C}(M) \mid \sup_{p \in M} |f(p)| < \infty \right\} \qquad (2.7.10)$$

on M. Show that they form a subalgebra of $\mathscr{C}(M)$ with unit which is closed under max, min, $|\cdot|$, and under complex conjugation.

(v) Define the supremum norm

$$\|f\|_\infty = \sup_{p \in M} |f(p)| \qquad (2.7.11)$$

for $f \in \mathscr{C}_b(M)$. Show that $\| \cdot \|$ is a norm on $\mathscr{C}_b(M)$.

(vi) Show that $\|fg\|_\infty \le \|f\|_\infty \|g\|_\infty$ for all $f, g \in \mathscr{C}_b(M)$ and find analogous estimates, equalities, or counter-examples for the pointwise maximum, minimum, the absolute value, and the complex conjugation instead of the product.

(vii) Show that $\mathscr{C}_b(M)$ is a complete normed space, i.e. a Banach space, with respect to the supremum norm. Which well-known theorem from elementary calculus is contained in this statement?

Together with the previous property, the completeness makes $\mathscr{C}_b(M)$ a *Banach algebra*, see also Definition 6.2.1.

Exercise 2.7.20 (Continuity is a local property) Consider a map $f \colon (M, \mathcal{M}) \longrightarrow (N, \mathcal{N})$ between topological spaces. Show that the following statements are equivalent:

(i) The map f is continuous.
(ii) For all subsets $A \subseteq M$ the maps $f|_A \colon (A, \mathcal{M}|_A) \longrightarrow (N, \mathcal{N})$ are continuous.
(iii) For all open covers $\{\mathcal{O}_i\}_{i \in I}$, i.e. $\mathcal{O}_i \in \mathcal{M}$ and $M = \bigcup_{i \in I} \mathcal{O}_i$, the restrictions $f|_{\mathcal{O}_i} \colon (\mathcal{O}_i, \mathcal{M}|_{\mathcal{O}_i}) \longrightarrow (N, \mathcal{N})$ are continuous for all $i \in I$.
(iv) There exists an open cover $\{\mathcal{O}_i\}_{i \in I}$ of M such that the restrictions $f|_{\mathcal{O}_i} \colon (\mathcal{O}_i, \mathcal{M}|_{\mathcal{O}_i}) \longrightarrow (N, \mathcal{N})$ are continuous for all $i \in I$.

Exercise 2.7.21 (Gluing of paths) Let $(M, \mathcal{M})$ be a topological space. Moreover, let $\gamma_1 \colon [a, b] \longrightarrow M$ and $\gamma_2 \colon [b, c] \longrightarrow M$ be continuous paths in M where $a < b < c$. Show that

$$\gamma(t) = \begin{cases} \gamma_1(t) & t \in [a, b] \\ \gamma_2(t) & t \in (b, c] \end{cases} \qquad (2.7.12)$$

yields a continuous path $\gamma \colon [a, c] \longrightarrow M$ provided $\gamma_1(b) = \gamma_2(b)$.

Hint: Suppose the converse and let $U \subseteq M$ be a neighbourhood of $\gamma(b) = \gamma_1(b) = \gamma_2(b)$ such that $\gamma^{-1}(V)$ is not a neighbourhood of $b \in [a, c]$. Consider then $\gamma_1^{-1}(V)$ and $\gamma_2^{-1}(V)$.

Exercise 2.7.22 (Connected components) Let $(M, \mathcal{M})$ be a topological space.

(i) Show that $p \sim q$ if $q \in \mathcal{C}(p)$ defines an equivalence relation on M. Hence M decomposes into mutually disjoint connected components.
   Hint: Use Proposition 2.5.9, (iii), to obtain the characterization that $\mathcal{C}(p)$ is the largest connected subset of M which contains p.

(ii) Show the analogous result for the path-connected components of M.

Exercise 2.7.23 (The topologist's sine curve) Let

$$S = \big\{(x, \sin(1/x)) \mid x \in (0, 1]\big\} \subseteq \mathbb{R}^2 \qquad (2.7.13)$$

be the graph of the function $x \mapsto \sin(1/x)$ defined on the interval $(0, 1]$, endowed with the subspace topology of $\mathbb{R}^2$.

(i) Sketch the graph S as well as its closure S^{cl} in $\mathbb{R}^2$. Which points are added when passing to the closure?

(ii) Show that S is connected and even path-connected.

(iii) Show that S^{cl} is connected but not path-connected.

(iv) Show that S^{cl} is not locally (path-) connected.

Exercise 2.7.24 (Cantor set I) Consider the following subsets of the closed unit interval $A_0 = [0, 1]$: in every new step one removes the open inner third of each piece, i.e. $A_1 = [0, \frac{1}{3}] \cup [\frac{2}{3}, 1]$, then $A_2 = [0, \frac{1}{9}] \cup [\frac{2}{9}, \frac{3}{9}] \cup [\frac{6}{9}, \frac{7}{9}] \cup [\frac{8}{9}, 1]$ etc., see Fig. 2.8. The Cantor set is then the infinite intersection

$$C = \bigcap_{n=0}^{\infty} A_n \subseteq [0, 1]. \qquad (2.7.14)$$

Show the following properties of C with respect to the subspace topology of $[0, 1]$:

(i) The Cantor set C is uncountable.
   Hint: Show that C is the set of those real numbers x in $[0, 1]$ which can be written as

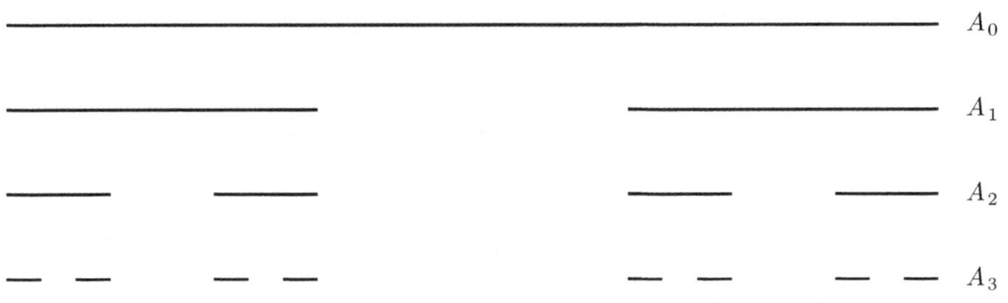

Fig. 2.8 The first iterations for the Cantor set

$$x = \sum_{n=1}^{\infty} \frac{a_n}{3^n} \qquad (2.7.15)$$

with a_n either 0 or 2.

(ii) The Cantor set C is closed.

(iii) The Cantor set C is nowhere dense.

(iv) The Cantor set C is totally disconnected.

(v) The Cantor set is not discrete.

(vi) For every $\epsilon > 0$ there is an $n \in \mathbb{N}$ and open intervals $I_1, \ldots, I_n$ with total length $|I_1| + \cdots + |I_n| < \epsilon$ such that

$$C \subseteq I_1 \cup \cdots \cup I_n. \qquad (2.7.16)$$

With other words, the Cantor set C has Lebesgue measure 0.

Exercise 2.7.25 (Separation properties I) Consider a set M with at least two elements endowed with the indiscrete topology. Which of the separation properties T_0, T_1, T_2, T_3, or T_4 are fulfilled? For which implications between the separation properties does this example provide counterexamples?

Exercise 2.7.26 (Separation properties II) Consider the set $M = \{1, 2, 3, 4\}$ with the topology $\mathcal{M} = \{\emptyset, \{1\}, \{1, 2\}, \{1, 3\}, \{1, 2, 3\}, M\}$.

(i) Show that $\mathcal{M}$ is indeed a topology.

(ii) Determine all closed subsets of $(M, \mathcal{M})$.

(iii) Which separation properties does $(M, \mathcal{M})$ fulfill?

Chapter 3
Construction of Topological Spaces

For a topological space $(M, \mathcal{M})$ we have already seen that any subset $N \subseteq M$ inherits a topology, the subspace topology $\mathcal{M}\big|_N$. This provides one important construction of topologies on certain sets. In this chapter we collect several further general constructions.

3.1 The Product Topology and Initial Topologies

Suppose $(M_i, \mathcal{M}_i)_{i \in I}$ is a family of topological spaces. Then we can consider the Cartesian product

$$M = \prod_{i \in I} M_i \tag{3.1.1}$$

of all the sets M_i, which we would like to endow now with a reasonable topology. One main motivation to consider Cartesian products is that one wants to treat many questions at once: all components M_i shall be handled simultaneously. Hence we would like to have a topology on M which is essentially designed in such a way that we can "do things componentwise". While this is clearly the heuristic idea, we have to be now more precise how one can achieve this goal. Since we are given the canonical projections

$$\mathrm{pr}_i : M \longrightarrow M_i \tag{3.1.2}$$

onto the i-th component for $i \in I$, we can use them to construct a topology on M. We are looking for a topology such that all the pr_i are continuous. Of course, the finest topology will always do the job as every subset of M is declared to be open: hence $\mathrm{pr}_i^{-1}(\mathcal{O}_i)$ is always open for any $\mathcal{O}_i \subseteq M$. Since this does not capture any particular structure and information about the topologies on the M_i, we want the topology on M to be *as coarse as possible*, preserving the continuity of the pr_i. In fact, as we shall see in the sequel, it will be a general theme that one constructs topologies by

© Springer International Publishing Switzerland 2014
S. Waldmann, *Topology*, DOI 10.1007/978-3-319-09680-3_3

requiring certain "extremal" properties like "the finest topology such that …" or "the coarsest topology such that …".

We can now realize the product topology explicitly as follows:

Definition 3.1.1 (*Product topology*) Let I be an index set and let $(M_i, \mathcal{M}_i)_{i \in I}$ be a family of topological spaces. Then on the Cartesian product

$$M = \prod_{i \in I} M_i \tag{3.1.3}$$

the product topology is determined by the subbasis

$$\mathcal{S} = \left\{ \mathrm{pr}_i^{-1}(\mathcal{O}_i) \ \middle| \ i \in I \text{ and } \mathcal{O}_i \in \mathcal{M}_i \right\}. \tag{3.1.4}$$

Before checking the continuity of the pr_i we note that a *basis* $\mathcal{B}$ for the product topology is obtained by taking finite intersections $\mathrm{pr}_{i_1}^{-1}(\mathcal{O}_{i_1}) \cap \cdots \cap \mathrm{pr}_{i_n}^{-1}(\mathcal{O}_{i_n})$ of subsets in $\mathcal{S}$. Note also that an arbitrary subset $U \subseteq M$ is in $\mathcal{S}$ iff $U = \prod_{i \in I} U_i$ with $U_i = M_i$ for all $i \in I$ except for one $i_0 \in I$ and $U_{i_0} \in \mathcal{M}_{i_0}$. Thus for a finite intersection U of such sets we still have $U_i = M_i$ for *all but finitely many $i \in I$*. Hence U is in $\mathcal{B}$ iff U is the product of U_i's with all U_i equal to M_i except for finitely many $U_{i_1} \in \mathcal{M}_{i_1}, \ldots, U_{i_n} \in \mathcal{M}_{i_n}$.

In the case of a finite Cartesian product $M = M_1 \times \cdots \times M_n$ we note that the above basis of the product is simply given by

$$\mathcal{B} = \left\{ \mathcal{O}_1 \times \cdots \times \mathcal{O}_n \ \middle| \ \mathcal{O}_1 \in \mathcal{M}_1, \ldots, \mathcal{O}_n \in \mathcal{M}_n \right\}, \tag{3.1.5}$$

which can be viewed as "open rectangles", see also Fig. 3.1. Clearly, the natural topology of $\mathbb{R}^n$ is the product topology of n copies of $\mathbb{R}$. For an infinite Cartesian product the idea of the open rectangles is misleading: most of the factors are equal to the whole space M_i and hence one should better think of "open cylinders" instead, see also Exercise 3.4.4 for a different topology on the Cartesian product.

We come now to the general characterization of the product topology which we took as the original motivation:

Proposition 3.1.2 *Let I be an index set and let $(M_i, \mathcal{M}_i)_{i \in I}$ be non-empty topological spaces. Endow the Cartesian product $M = \prod_{i \in I} M_i$ with the product topology.*

(i) *For every $i \in I$, the projection $\mathrm{pr}_i : M \longrightarrow M_i$ is continuous and open.*

(ii) *The product topology is the coarsest topology on M such that all the projections are continuous.*

(iii) *For another topological space $(N, \mathcal{N})$ a map $f : N \longrightarrow M$ is continuous iff $\mathrm{pr}_i \circ f : N \longrightarrow M_i$ is continuous for all $i \in I$.*

(iv) *The Cartesian product topology is T_k for $k \in \{0, 1, 2, 3\}$ iff each factor $(M_i, \mathcal{M}_i)$ is T_k.*

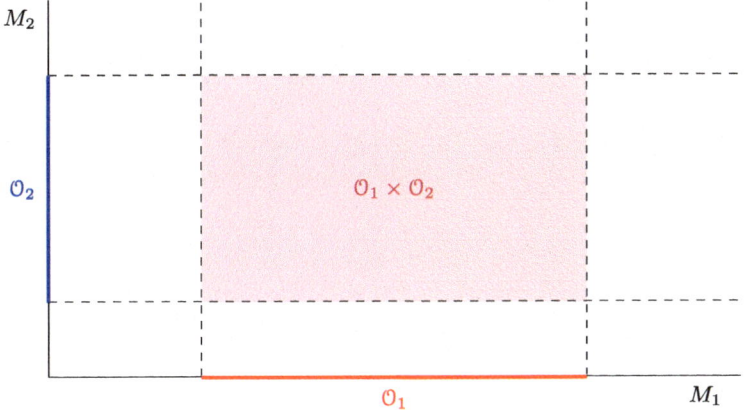

Fig. 3.1 Open rectangles in the Cartesian product

Proof The continuity of the pr_i is clear by the very definition of the subbasis (3.1.4), also the second part follows directly from this since any topology with all pr_i continuous has to contain the subsets $\mathrm{pr}_i^{-1}(\mathcal{O}_i)$ with $\mathcal{O}_i \in \mathcal{M}_i$ and $i \in I$. Now let $U = \mathrm{pr}_{i_1}^{-1}(\mathcal{O}_{i_1}) \cap \cdots \cap \mathrm{pr}_{i_n}^{-1}(\mathcal{O}_{i_n}) \in \mathcal{B}$ be an element of the basis $\mathcal{B}$ defined by the subbasis $\mathcal{S}$ as above, where $\mathcal{O}_{i_1} \in \mathcal{M}_{i_1}, \ldots, \mathcal{O}_{i_n} \in \mathcal{M}_{i_n}$. Then $\mathrm{pr}_i(U)$ is either M_i if i is different from $i_1, \ldots, i_n$, or given by $\mathcal{O}_{i_k}$ if $i = i_k$. In both cases, the result is an open subset of M_i. Thus pr_i is an open map as it suffices to check this on a basis of the topology, see Exercise 2.7.17. For the third part it is clear that $\mathrm{pr}_i \circ f$ is continuous if f is continuous. Thus assume $\mathrm{pr}_i \circ f$ is continuous for all $i \in I$. Let $U \subseteq M$ be in the subbasis $\mathcal{S}$, i.e. $U = \mathrm{pr}_{i_0}^{-1}(\mathcal{O}_{i_0})$ for some $i_0 \in I$. Then $f^{-1}(\mathrm{pr}_{i_0}^{-1}(\mathcal{O}_{i_0})) = (\mathrm{pr}_{i_0} \circ f)^{-1}(\mathcal{O}_{i_0}) \in \mathcal{N}$ by the continuity of $\mathrm{pr}_{i_0} \circ f$. From Proposition 2.4.2, (iv), we know that it is enough to test continuity on a subbasis of the target. For the last part, suppose first that the Cartesian product topology is T_k and let $i_0 \in I$ be fixed. For every $i \neq i_0$ we choose a point $p_i \in M_i$ and define the map

$$s_{i_0} \colon p \in M_{i_0} \;\mapsto\; (p_j)_{j \in I} \in M,$$

where $p_{j_0} = p$ and for $j \neq i_0$ we take the fixed choices p_i. Note that here we need the Axiom of Choice in order to find a choice of points $p_i \in M_i$ for all $i \neq i_0$. Clearly, $\mathrm{pr}_{i_0} \circ s_{i_0} = \mathrm{id}_{M_{i_0}}$. A map s_{i_0} with this property is also called a continuous *section* of pr_{i_0}. The section property of s_{i_0} shows that pr_{i_0} restricted to $s_{i_0}(M_{i_0}) \subseteq M$ is a bijection onto M_{i_0}. Since the map $\mathrm{pr}_{i_0}\big|_{s_{i_0}(M_{i_0})}$ is continuous and still open, it is a homeomorphism. It follows that M_{i_0} is homeomorphic to a subspace $s_{i_0}(M_{i_0})$ of the Cartesian product M. Hence by Proposition 2.6.12 the factor M_{i_0} is T_k, too. The reverse implication is part of Exercise 3.4.1. $\qquad\square$

The crucial point for the construction of the Cartesian product topology is the required continuity of the projection maps $\mathrm{pr}_i \colon M \longrightarrow M_i$. We can consider this in some larger generality as follows: Let M be an arbitrary set and let

$$p_i : M \longrightarrow M_i \tag{3.1.6}$$

be maps into topological spaces $(M_i, \mathcal{M}_i)_{i \in I}$ where I is some index set. We can now repeat the construction of the product topology even though M and the maps p_i are arbitrary:

Proposition 3.1.3 *Let M be a set, let $(M_i, \mathcal{M}_i)_{i \in I}$ be topological spaces with an index set I, and let $p_i : M \longrightarrow M_i$ be maps.*

(i) The collection

$$\mathcal{S} = \left\{ p_i^{-1}(\mathcal{O}_i) \subseteq M \mid i \in I, \mathcal{O}_i \in \mathcal{M}_i \right\} \tag{3.1.7}$$

is a subbasis of the coarsest topology $\mathcal{M}$ on M such that all the maps p_i are continuous.

(ii) This topology $\mathcal{M}$ is the unique topology on M with the universal property that for every topological space $(N, \mathcal{N})$ and every map $f : N \longrightarrow M$ we have: f is continuous iff the compositions $p_i \circ f : N \longrightarrow M_i$ are continuous for all $i \in I$.

Proof The first part is completely analogous to the corresponding statements in Proposition 3.1.2. Also the universal property of $\mathcal{M}$ follows analogously to the case of the product topology. Thus it remains to check uniqueness: let $\mathcal{M}'$ be another topology on M with the universal property as in (ii). Then consider the identity map

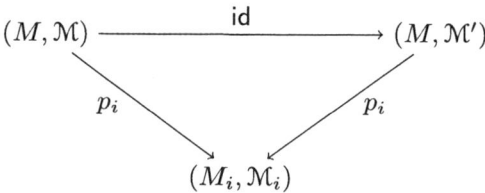

of M which is continuous by the universal property of $\mathcal{M}'$. But we can exchange $\mathcal{M}'$ by $\mathcal{M}$ in (i) showing that also id: $(M, \mathcal{M}') \longrightarrow (M, \mathcal{M})$ is continuous. Thus $\mathcal{M} = \mathcal{M}'$ follows. $\square$

Definition 3.1.4 (*Initial topology*) The topology $\mathcal{M}$ on M described in Proposition 3.1.3 is called the initial topology with respect to the maps $p_i : M \longrightarrow M_i$.

The argument concerning the uniqueness of the initial topology can be used at many other places where certain objects are characterized by universal properties in categories. We do not formulate this precisely but we come back to it from time to time in examples.

Remark 3.1.5 With this notion, the product topology on $M = \prod_{i \in I} M_i$ is the initial topology with respect to all the projection maps $\mathrm{pr}_i : M \longrightarrow M_i$. In particular, the product topology is the unique topology with the property of Proposition 3.1.2, (iii).

Example 3.1.6 Let $(M, \mathcal{M})$ be a topological space an let $N \subseteq M$. Then the subspace topology $\mathcal{M}|_N$ of N is the initial topology with respect to the natural inclusion map $\iota : N \longrightarrow M$.

3.2 Final Topologies and Quotients

In some sense dual to the construction of the initial topology is the final topology: here we reverse all arrows and want a similar characterization. Thus let M be a given set and let again $(M_i, \mathcal{M}_i)_{i \in I}$ be a family of topological spaces indexed by some index set I. Moreover, let

$$q_i : M_i \longrightarrow M \tag{3.2.1}$$

be given maps. Looking for the coarsest topology $\mathcal{M}$ on M such that all the q_i become continuous is now the wrong question as we can always take the indiscrete topology. However, asking for the finest topology with this continuity property becomes interesting. Alternatively, it is characterized by a universal property:

Proposition 3.2.1 *Let M be a set and let $q_i : M_i \longrightarrow M$ be maps from topological spaces $(M_i, \mathcal{M}_i)_{i \in I}$ into M for some index set I.*

(i) *On M there exists a unique topology $\mathcal{M}$ with the universal property that a map $f : M \longrightarrow N$ into a topological space $(N, \mathcal{N})$ is continuous iff $f \circ q_i : M_i \longrightarrow N$ is continuous for all $i \in I$.*

(ii) *This topology $\mathcal{M}$ is the finest topology on M such that all the maps $q_i : M_i \longrightarrow M$ are continuous. A subset $\mathcal{O} \subseteq M$ is open iff $q_i^{-1}(\mathcal{O}) \subseteq M_i$ is open for all $i \in I$.*

Proof The proof is completely analogous to the one of Proposition 3.1.3. First we show uniqueness. Thus let $\mathcal{M}$ and $\mathcal{M}'$ be two topologies on M with the above universal property. Then $\mathrm{id} : (M, \mathcal{M}') \longrightarrow (M, \mathcal{M}')$ is continuous and hence $q_i : (M_i, \mathcal{M}_i) \longrightarrow (M, \mathcal{M}')$ is continuous. But then also $\mathrm{id} : (M, \mathcal{M}) \longrightarrow (M, \mathcal{M}')$ is continuous by the universal property of $\mathcal{M}$. This means $\mathcal{M}' \subseteq \mathcal{M}$ and by symmetry we have $\mathcal{M}' = \mathcal{M}$. For the existence define $\mathcal{M}$ to be the topology by declaring $\mathcal{O} \subseteq M$ to be open if $q_i^{-1}(\mathcal{O})$ is open in M_i for all $i \in I$. Since taking preimages is compatible with arbitrary unions and intersections it follows that $\mathcal{M}$ is indeed a topology. Clearly, it is the finest topology such that all the $q_i : M_i \longrightarrow M$ are continuous. Now let $f : M \longrightarrow N$ be given such that all $f \circ q_i$ are continuous and let $\mathcal{O} \subseteq N$ be open. Then $f^{-1}(\mathcal{O}) \subseteq M$ is open since $q_i^{-1}(f^{-1}(\mathcal{O})) = (f \circ q_i)^{-1}(\mathcal{O}) \subseteq M_i$ is open for all $i \in I$ thanks to the continuity of $f \circ q_i$. Thus f is continuous. Conversely, if f is continuous then also $f \circ q_i$ is continuous for all $i \in I$ since the q_i are continuous. Hence $\mathcal{M}$ has the universal property. $\qquad\square$

Definition 3.2.2 (*Final topology*) The topology $\mathcal{M}$ on M described in Proposition 3.2.1 is called the final topology with respect to the maps $q_i : M_i \longrightarrow M$.

The final topology often arises in the construction of *quotients*. Let N be an arbitrary set with an equivalence relation $\sim$ and consider the quotient set

$$M = N/\!\sim, \tag{3.2.2}$$

i.e. the set of equivalence classes. Mapping a point $p \in N$ to its equivalence class $[p] \in M$ gives a canonical quotient map

$$\pi : N \longrightarrow M = N/\!\sim. \tag{3.2.3}$$

Definition 3.2.3 (*Quotient topology*) Let $(N, \mathcal{N})$ be a topological space with an equivalence relation $\sim$. Then the final topology on $M = N/\!\sim$ with respect to the quotient map $\pi : N \longrightarrow M$ is called the quotient topology.

Remark 3.2.4 Unlike the initial topology the final topology and in particular the quotient topology does not respect the separation axioms very well: there are easy examples of non-Hausdorff quotients even though the original space is Hausdorff.

Example 3.2.5 Let $N \subseteq \mathbb{R}^2$ be given by two copies of the open unit interval $(0, 1)$ sitting at $y = 0$ and $y = 1$, i.e.

$$N = \left\{ (x, y) \mid x \in (0, 1), y \in \{0, 1\} \right\} \subseteq \mathbb{R}^2. \tag{3.2.4}$$

This is clearly a Hausdorff space having two connected components. We define now $(x, y) \sim (x', y')$ if either $x = x' \in (0, \frac{1}{2})$ and $y, y' \in \{0, 1\}$ or if $x = x' \in [\frac{1}{2}, 1)$ and $y = y'$. Geometrically, this means to glue the open intervals $(0, \frac{1}{2}) \times \{0\}$ and $(0, \frac{1}{2}) \times \{1\}$ together to obtain an interval $(0, \frac{1}{2})$ and attach the two half-closed intervals $[\frac{1}{2}, 1) \times \{0\}$ and $[\frac{1}{2}, 1) \times \{1\}$ to it. The resulting quotient M looks like a tripod but with a *doubled* center consisting of the two equivalence classes $[(\frac{1}{2}, 0)]$ and $[(\frac{1}{2}, 1)]$, see Fig. 3.2. It is now easy to see that at this double point, the Hausdorff property fails.

However, there is a very simple and useful criterion whether the quotient topology is Hausdorff, provided the quotient map is an open map: unfortunately this needs not to be true, see Exercise 3.4.3. Nevertheless, under this additional assumption one gets the following result:

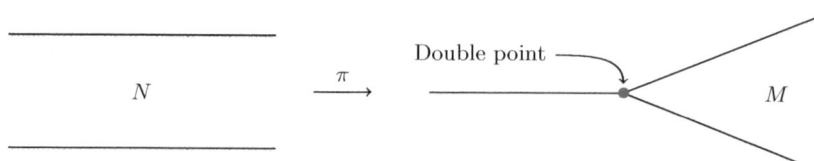

Fig. 3.2 The tripod with a double point in the center

Proposition 3.2.6 *Let $\sim$ be an equivalence relation on a topological space $(M, \mathcal{M})$ and consider the quotient $\pi \colon M \longrightarrow M/\sim$.*

(i) If the quotient $M/\sim$ is Hausdorff then the relation $\sim$, viewed as a subset of the Cartesian product $M \times M$ is closed.

(ii) If the quotient map is open and $\sim$ is a closed subset of $M \times M$ then the quotient $M/\sim$ is Hausdorff.

Proof We know from Exercise 3.4.2 that $M/\sim$ is Hausdorff iff the diagonal $\Delta_{M/\sim} \subseteq (M/\sim) \times (M/\sim)$ is closed. Now assume that the quotient is Hausdorff, then also $(\pi \times \pi)^{-1}(\Delta_{M/\sim}) \subseteq M \times M$ is closed since $\pi \times \pi$ is continuous. We have

$$
\begin{aligned}
(\pi \times \pi)^{-1}(\Delta_{M/\sim}) &= \big\{ (p, q) \in M \times M \mid \pi(p) = \pi(q) \big\} \\
&= \big\{ (p, q) \in M \times M \mid p \sim q \big\} \\
&= \sim,
\end{aligned}
$$

showing the first part. For the second, let $\pi(p) \neq \pi(q)$ be two different points in $M/\sim$. Fix a pair of representatives p and q then $(p, q) \notin \sim$. Since $\sim$ is closed, $M \times M \setminus \sim$ is open and hence we find open neighbourhoods $U \subseteq M$ of p and $V \subseteq M$ of q with $(U \times V) \cap \sim = \emptyset$. Since the quotient map π is open by assumption, $\pi(U)$ and $\pi(V)$ are open neighbourhoods of $\pi(p)$ and $\pi(q)$, respectively. We claim that $\pi(U) \cap \pi(V) = \emptyset$. Indeed, assume that $\pi(z) \in \pi(U) \cap \pi(V)$ then $z \in \pi^{-1}(\pi(U) \cap \pi(V)) = \pi^{-1}(\pi(U)) \cap \pi^{-1}(\pi(V))$. Now $z \in \pi^{-1}(\pi(U))$ means $z \sim x$ with some $x \in U$. Analogously, we get $z \sim y$ for some $y \in V$. But then $x \sim y$ follows by the transitivity of $\sim$, contradicting $(U \times V) \cap \sim = \emptyset$. $\square$

Example 3.2.7 (Group action) Let G be a group and let M be a set. Recall that a *group action* is a map $\Phi \colon G \times M \longrightarrow M$ such that

$$\Phi(e, p) = p \quad \text{and} \quad \Phi(g, \Phi(h, p)) = \Phi(gh, p) \tag{3.2.5}$$

for all $p \in M$ and $g, h \in G$. We write $\Phi_g(p) = \Phi(g, p)$ in order to view Φ_g as a map $\Phi_g \colon M \longrightarrow M$. Clearly, this is a bijection with $\Phi_g^{-1} = \Phi_{g^{-1}}$ and the map $g \mapsto \Phi_g$ gives a group homomorphism from G into the permutation group of M. Now if $(M, \mathcal{M})$ is in addition a topological space we say that the group action Φ is by homeomorphisms if all Φ_g are homeomorphisms. Note that this is equivalent to require that Φ_g is continuous for all $g \in G$. The *orbit* of $p \in M$ is the collection of all the points

$$G \cdot p = \big\{ \Phi_g(p) \mid g \in G \big\} \subseteq M, \tag{3.2.6}$$

which can be reached from p by applying a group element to it. This defines an equivalence relation by $p \sim q$ if $p \in G \cdot q$ which means that p and q are in the same orbit. Then the quotient

$$\pi \colon M \longrightarrow M/G = M/\sim, \tag{3.2.7}$$

endowed with the quotient topology is called the *orbit space* of the group action. Many important examples of topological spaces can be viewed as being orbit spaces for certain group action, see Exercise 3.4.7. Moreover, quotients by group actions have additional nice properties, see Exercise 3.4.8.

3.3 Topological Manifolds

In this subsection we give a very brief introduction to an important class of topological spaces, the topological manifolds. The basic idea is very simple: we want a topological space which looks *locally* like an open subset of $\mathbb{R}^n$.

Thus let $(M, \mathcal{M})$ be a topological space and let $U \subseteq M$ be an open subset. A homeomorphism

$$x: U \longrightarrow x(U) \subseteq \mathbb{R}^n \tag{3.3.1}$$

onto an open subset $x(U) \subseteq \mathbb{R}^n$ is called an n-*dimensional local chart* for $(M, \mathcal{M})$. The numbers $x(p) = (x^1(p), \ldots, x^n(p)) \in \mathbb{R}^n$ for $p \in U$ are called the *local coordinates* of $p \in M$ with respect to the local chart (U, x). If we can find such a local chart around every point in M then we get a topological manifold. For various reasons it is useful to require two additional properties of $(M, \mathcal{M})$. We want the topology to be second countable and Hausdorff:

Definition 3.3.1 (*Topological manifold*) Let $(M, \mathcal{M})$ be a topological space. Then $(M, \mathcal{M})$ is called a topological manifold of dimension $n \in \mathbb{N}_0$ if the following three properties are fulfilled:

(i) For every $p \in M$ there exists an n-dimensional local chart (U, x) of $(M, \mathcal{M})$ with $p \in U$.
(ii) The topology $\mathcal{M}$ is second countable.
(iii) The topology $\mathcal{M}$ is Hausdorff.

Remark 3.3.2 It is clear from the definition that a topological manifold is locally path-connected since the open balls in $\mathbb{R}^n$ are locally path-connected. Hence for topological manifolds the notions of connectedness and path-connectedness coincide thanks to Proposition 2.5.12 and connected components are both closed and open subsets.

We introduce some more notation. For a general topological space $(M, \mathcal{M})$ an *open cover* $\{U_i\}_{i \in I}$ is a collection of open subsets $U_i \in \mathcal{M}$ with

$$M = \bigcup_{i \in I} U_i. \tag{3.3.2}$$

Analogously, we define an open cover of a subset $N \subseteq M$. The cover is called *finite* or *countable* if I is finite or countable, respectively. For a topological manifold

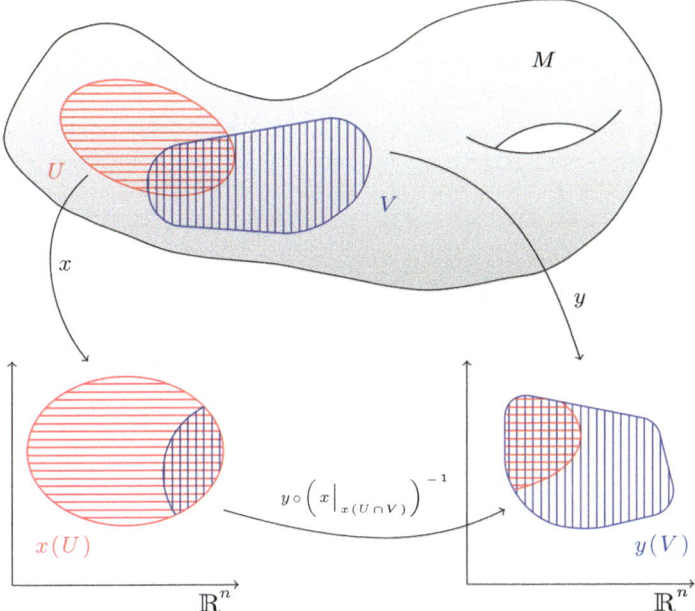

Fig. 3.3 Change of coordinates

we need an open cover by local charts $(U_i, x_i)_{i \in I}$. Such a cover is also called a *topological atlas*. Clearly, this notion still makes sense for $(M, \mathcal{M})$ not necessarily satisfying the Hausdorff property and the second countable property. Suppose now that (U, x) and (V, y) are n-dimensional local charts of $(M, \mathcal{M})$ with $U \cap V \neq \emptyset$. Since $x \colon U \longrightarrow x(U)$ is a homeomorphism, also $x|_{U \cap V} \colon U \cap V \longrightarrow x(U \cap V)$ is a homeomorphism, the same holds for $y|_{U \cap V} \colon U \cap V \longrightarrow y(U \cap V)$. This implies that

$$y \circ \left(x|_{U \cap V} \right)^{-1} \colon x(U \cap V) \longrightarrow y(U \cap V) \tag{3.3.3}$$

is a homeomorphism between two open subsets of $\mathbb{R}^n$. We call this map the *change of coordinates* from the chart (U, x) to the chart (V, y), see Fig. 3.3.

Now if we have a topological atlas $(U_i, x_i)_{i \in I}$ we get changes of coordinates

$$\varphi_{ij} = x_i \circ \left(x_j|_{U_i \cap U_j} \right)^{-1} \colon x_j(U_i \cap U_j) \longrightarrow x_i(U_i \cap U_j) \tag{3.3.4}$$

for all pairs i, j with $U_i \cap U_j \neq \emptyset$.

Lemma 3.3.3 (Cocycle property) *For an n-dimensional topological atlas $(U_i, x_i)_{i \in I}$ of a topological space $(M, \mathcal{M})$ the changes of coordinates $\{\varphi_{ij}\}_{i,j \in I}$ satisfy on $x_j(U_i \cap U_j)$*

$$\varphi_{ij} = \varphi_{ji}^{-1} \quad and \quad \varphi_{ik} \circ \varphi_{kj} \circ \varphi_{ji} = \mathsf{id} \tag{3.3.5}$$

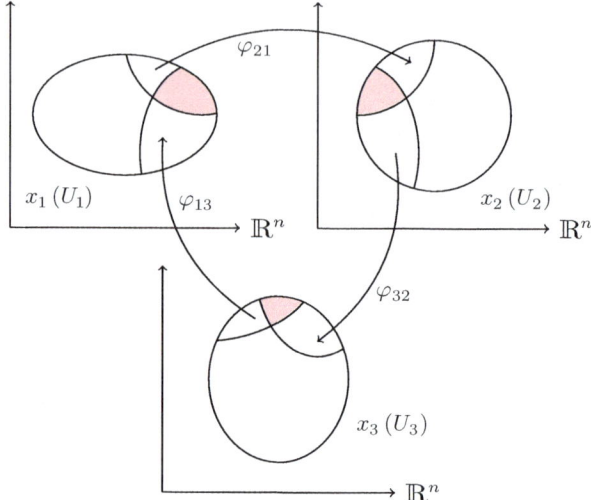

Fig. 3.4 The cocycle identity: moving points in the grey area around yields the identity

on $x_i(U_i \cap U_j \cap U_k)$ whenever $U_i \cap U_j \neq \emptyset$ and $U_i \cap U_j \cap U_k \neq \emptyset$, respectively.

Proof After unwinding the definition, this is completely obvious, see also Fig. 3.4.   □

The important point is that the relation (3.3.5) is entirely evaluated in $\mathbb{R}^n$ and does no longer refer to M itself. This suggest to build topological manifolds by patching together local charts provided the changes of coordinates satisfy the cocycle condition. This is indeed possible by the following construction: Let I be an index set and let $V_i \subseteq \mathbb{R}^n$ be non-empty and open for $i \in I$. Moreover, let $V_{ij} \subseteq V_j$ be an open subset for every pair $i \neq j$. Next, if $V_{ij} \neq \emptyset$, assume we are given a homeomorphism

$$\varphi_{ij} \colon V_{ij} \longrightarrow V_{ji} \tag{3.3.6}$$

In particular, we require $V_{ij} \neq \emptyset$ iff $V_{ji} \neq \emptyset$. It will be useful to set $V_{ii} = V_i$ and $\varphi_{ii} = \mathrm{id}_{V_i}$ for all $i \in I$. We want to use the homeomorphisms φ_{ij} to glue the coordinate patches together by identifying $V_{ij} \subseteq V_j$ with $V_{ji} \subseteq V_i$. To formulate the cocycle condition we thus have to require that the intersection $V_{ij} \cap V_{kj} \subseteq V_{ij}$ is mapped homeomorphically to $V_{ji} \cap V_{ki} \subseteq V_{ji}$ under φ_{ij} whenever the intersection is non-empty.

Lemma 3.3.4 *Suppose we are given non-empty open $V_i \subseteq \mathbb{R}^n$, open $V_{ij} \subseteq V_j$, and homeomorphisms $\varphi_{ij} \colon V_{ij} \longrightarrow V_{ji}$ for $V_{ij} \neq \emptyset$ such that*

$$\varphi_{ij} \colon V_{ij} \cap V_{kj} \longrightarrow V_{ji} \cap V_{ki} \tag{3.3.7}$$

is a homeomorphism whenever the intersection is non-empty. If in addition $\varphi_{ij} = \varphi_{ji}^{-1}$ and $\varphi_{ik} \circ \varphi_{kj} \circ \varphi_{ji}|_{V_{ji} \cap V_{ki}} = \mathrm{id}$ whenever $V_{ki} \cap V_{ji} \subseteq V_i$ is non-empty, then we get an equivalence relation $\sim$ on the disjoint union

$$\tilde{M} = \coprod_{i \in I} V_i \tag{3.3.8}$$

via setting $(p, i) \sim (q, j)$ if either $p = q$ and $i = j$ or $p \in V_{ji}$ with $q = \varphi_{ji}(p)$.

Proof Clearly (p, i) is equivalent to itself. From $\varphi_{ij} = \varphi_{ji}^{-1}$ we get symmetry and the cocycle condition $\varphi_{ik} \circ \varphi_{ki} \circ \varphi_{ji} = \mathrm{id}$ gives transitivity. □

The main point is now that $\tilde{M}$ is canonically a topological space with a topological atlas: we have the canonical inclusion maps

$$\varphi_i \colon V_i \longrightarrow \tilde{M}, \tag{3.3.9}$$

from which we can get the final topology on $\tilde{M}$. Being the disjoint union this means that $\mathcal{O} \subseteq \tilde{M}$ is open iff $\varphi_i^{-1}(\mathcal{O}) = \mathcal{O} \cap V_i$ is open for all $i \in I$. With other words, $\mathcal{O}$ is the disjoint union of open subsets $\mathcal{O}_i \subseteq V_i$ for each $i \in I$. This construction works also in general and is called *disconnected sum*, see also Exercise 3.4.5. Now the $\{V_i\}_{i \in I}$, viewed as open subsets of $\tilde{M}$, form an open cover and hence we get a topological atlas by taking just the identity map $V_i \subseteq \mathbb{R}^n$ as local coordinates on the chart V_i.

Lemma 3.3.5 *The topological space $\tilde{M}$ is Hausdorff. Moreover, it is second countable if I is at most countable.*

Proof Since for different $i \neq j$ we have $V_i \cap V_j = \emptyset$ inside $\tilde{M}$, the Hausdorff property can be decided inside a single V_i. There it is fulfilled as $V_i \subseteq \mathbb{R}^n$ is clearly Hausdorff. Since each V_i is second countable also their disconnected sum $\tilde{M}$ is second countable as long as I is at most countable. □

The last step of the construction consists in passing to the quotient

$$\pi \colon \tilde{M} \longrightarrow M = \tilde{M}/\!\!\sim \tag{3.3.10}$$

with respect to the equivalence relation $\sim$ from Lemma 3.3.4. We endow M with the quotient topology, i.e. $U \subseteq M$ is open if $\pi^{-1}(U) \subseteq \tilde{M}$ is open. We consider now the images of the open subsets $V_i \subseteq \tilde{M}$ under π. Here we have the following crucial observation:

Lemma 3.3.6 *Let $U_i = \pi(V_i) \subseteq M$ for $i \in I$. Then*

$$\pi\big|_{V_i} \colon V_i \longrightarrow U_i \tag{3.3.11}$$

is a homeomorphism.

Proof By definition, π is surjective. From the definition of the equivalence relation we see that π is also injective when restricted to a single V_i. Hence we have a continuous bijection of which we have to show that $(\pi|_{V_i})^{-1}$ is continuous as well. Let $(p, i) \in V_i$ and consider the equivalence class $\pi(p, i) \in M$. Then $\pi^{-1}(\pi(p, i))$ consists of all points (q, j) in $\tilde{M}$ which are equivalent to (p, i). Hence

$$\pi^{-1}(\pi((p, i))) = \left\{ (q, j) \mid \text{either } p = q \text{ and } i = j \text{ or } q = \varphi_{ji}(p) \right\}.$$

If $\mathcal{O} \subseteq V_i$ is open then $\pi^{-1}(\pi(\mathcal{O}))$ consists of all points in $\pi^{-1}(\pi((p, i)))$ for all $p \in \mathcal{O}$. We get these points either for $j = i$ and hence $\mathcal{O} \subseteq \pi^{-1}(\pi(\mathcal{O}))$ or for $i \neq j$ and $p \in V_{ji} \cap \mathcal{O}$ with $q = \varphi_{ji}(p) \in \pi^{-1}(\pi(\mathcal{O}))$. Thus

$$\pi^{-1}(\pi(\mathcal{O})) = \bigcup_{j \in I \setminus \{i\}} \varphi_{ji}(V_{ji} \cap \mathcal{O}) \cup \mathcal{O},$$

where the unions are disjoint as we are in different connected components of $\tilde{M}$. Since $V_{ji} \subseteq V_i$ is open and since φ_{ji} is a homeomorphism, also $\varphi_{ji}(U_{ji} \cap \mathcal{O}) \subseteq V_j$ is open for all $j \in I \setminus \{i\}$. Thus $\pi^{-1}(\pi(\mathcal{O}))$ is open showing that π is an *open* mapping. By Proposition 2.4.8, (ii), it follows that $\pi|_{V_i}$ is a homeomorphism, since $\pi|_{V_i}$ is still an open mapping according to $V_i \subseteq \tilde{M}$ being open itself.   □

Thus we can consider the (continuous) inverse of $\pi|_{V_i}$ denoted by

$$x_i = \left(\pi|_{V_i}\right)^{-1} \colon U_i \longrightarrow V_i, \tag{3.3.12}$$

which gives an n-dimensional local chart for M. Of course, the open subsets $U_i \subseteq M$ cover all of M by the very construction and thus we get a topological atlas for M. We summarise this construction:

Theorem 3.3.7 (Construction of topological manifolds) *Let I be an index set and let $V_i \subseteq \mathbb{R}^n$ be non-empty open subsets for $i \in I$. For $i \neq j$, let $V_{ij} \subseteq V_j$ be an open (possible empty) subset together with a homeomorphism $\varphi_{ij} \colon V_{ij} \longrightarrow V_{ji}$ for those (i, j) with $V_{ij} \neq \emptyset$. For $i = j$ we set $V_{ii} = V_i$ and $\varphi_{ii} = \mathsf{id}_{V_i}$. Suppose that $\varphi_{ij} \colon V_{ij} \cap V_{kj} \longrightarrow V_{ji} \cap V_{ki}$ is a homeomorphism whenever the intersection is non-empty and*

$$\varphi_{ij} = \varphi_{ji}^{-1} \quad \text{and} \quad \varphi_{ik} \circ \varphi_{kj} \circ \varphi_{jk} = \mathsf{id} \tag{3.3.13}$$

on non-trivial intersections. Endow the disconnected topological sum $\tilde{M} = \coprod_{i \in I} V_i$ with the equivalence relation

$$p \sim q \quad \text{if} \quad p \in V_i, q \in V_j, q = \varphi_{ji}(p). \tag{3.3.14}$$

Then the quotient $\pi \colon \tilde{M} \longrightarrow \tilde{M}/\!\sim \, = M$ has a topological atlas $(U_i, x_i)_{i \in I}$ given by

$$U_i = \pi(V_i) \quad and \quad x_i = \left(\pi\big|_{V_i}\right)^{-1}. \qquad (3.3.15)$$

The quotient is second countable if I is at most countable.

Thus we obtain almost a topological manifold: the only thing missing is the Hausdorff property. In view of Example 3.2.5 this can indeed *fail*. We have to guarantee the Hausdorff property by a separate investigation, sometimes by some rather ad-hoc arguments.

We conclude this section with an example:

Example 3.3.8 (Spheres) We consider the 2-sphere $\mathbb{S}^2 \subseteq \mathbb{R}^3$ with its usual subspace topology. Denote the north pole $(0, 0, 1)$ by N and the south pole $(0, 0, -1)$ by S. Denote by $x_N \colon \mathbb{S}^2 \setminus \{N\} \longrightarrow \mathbb{R}^2 \subseteq \mathbb{R}^3$ the stereographic projection onto the xy-plane from the north pole. Analogously, we can project from the south pole yielding $x_S \colon \mathbb{S}^2 \setminus \{S\} \longrightarrow \mathbb{R}^2 \subseteq \mathbb{R}^3$. It is now easy to see that these two maps provide a topological atlas for $\mathbb{S}^2$. Analogous projections are possible for $\mathbb{S}^n$ for all $n \in \mathbb{N}$ showing that all spheres are topological manifolds, see also Exercise 3.4.6.

Of course the story does not end here with topological manifolds. Instead, one uses a topological atlas to require more: the changes of coordinates should have some better regularity than just being continuous. One requires them to be e.g. $\mathscr{C}^k$-*diffeomorphisms* for some $k \in \mathbb{N} \cup \{\infty\}$. This makes sense since for open subsets of $\mathbb{R}^n$ we know what differentiability of functions means. Another option is to require *real-analytic* changes of coordinates, i.e. $\mathscr{C}^\omega$-diffeomorphisms. Finally, we can also consider the case of $2n$-dimensional local charts, identify $\mathbb{R}^{2n} \cong \mathbb{C}^n$, and require the changes of coordinates to be *holomorphic*. This leads to various notions of manifolds: $\mathscr{C}^k$-manifolds, real-analytic manifolds, and complex manifolds. There is an abundance of literature on manifolds, see e.g. [4, 6, 12, 21–23, 25] to name just a few.

3.4 Exercises

Exercise 3.4.1 (Separation properties for products) Let $(M_i, \mathcal{M}_i)_{i \in I}$ be topological space satisfying the separation property T_k, where $k \in \{0, 1, 2, 3\}$. Endow $M = \prod_{i \in I} M_i$ with the product topology. Show that M satisfies T_k as well.
Hint: Consider first the (most difficult) case of T_3 and use the characterization from Proposition 2.6.5.

Exercise 3.4.2 (The diagonal) Let $(M, \mathcal{M})$ be a topological space. Consider the Cartesian product $M \times M$ with its product topology. Define the diagonal map

$$\Delta \colon M \ni p \mapsto (p, p) \in M \times M. \qquad (3.4.1)$$

(i) Let $X, Y \subseteq M$ be subsets. Show that

$$\Delta^{-1}(X \times Y) = X \cap Y. \qquad (3.4.2)$$

 (ii) Show that Δ is injective and continuous.
(iii) Show that the diagonal map is an embedding.
       Hint: Determine the subspace topology of $\Delta(M) \subseteq M \times M$ explicitly using (ii).
 (iv) Suppose that now M is in addition Hausdorff. Show that the image $\Delta_M = \Delta(M)$
       of M under the diagonal map is a closed subset of $M \times M$.
  (v) Show that conversely, a closed diagonal $\Delta_M \subseteq M \times M$ implies that M is
       Hausdorff.

Thus for a Hausdorff topological space the diagonal is an embedding with closed
image, as the graphical description would suggest.

Exercise 3.4.3 (A non-open quotient map) Consider the subset

$$M = \mathbb{R} \cup \{p\} \subseteq \mathbb{R}^2 \tag{3.4.3}$$

where p is the point $(0, 1)$. Define an equivalence relation on M by the only non-
trivial relation $(0, 0) \sim (0, 1)$.

 (i) Show that the quotient $M/\!\!\sim$ is homeomorphic to $\mathbb{R}$ by the obvious identifica-
      tion.
(ii) Show that the quotient map $\pi \colon M \longrightarrow M/\!\!\sim$ is not an open map.

Exercise 3.4.4 (A finer topology on the Cartesian product) Let $(M_i, \mathcal{M}_i)_{i \in I}$ be a
non-empty set of topological spaces. On their Cartesian product $M = \prod_{i \in I} M_i$ one
defines the following topology: let

$$\mathcal{S} = \left\{ \mathcal{O} = \prod_{i \in I} \mathcal{O}_i \;\middle|\; \mathcal{O}_i \in \mathcal{M}_i \text{ for all } i \in I \right\} \tag{3.4.4}$$

and define $\mathcal{M}$ to be the topology generated by the subbasis $\mathcal{S}$.

  (i) Show that $\mathcal{S}$ is not a basis of a topology but only a subbasis in general.
 (ii) Show that the topology $\mathcal{M}$ coincides with the Cartesian product topology if the
       index set I is finite.
(iii) Show that in general the topology $\mathcal{M}$ is finer than the Cartesian product topology
       and give an example where it is *strictly finer*.

This topology is sometimes named the *box topology* for the Cartesian product.

Exercise 3.4.5 (The disconnected sum) Let $(M_i, \mathcal{M}_i)_{i \in I}$ be a set of topological
spaces. Then we consider their disjoint union $M = \coprod_{i \in I} M_i$ and define $\mathcal{O} \subseteq M$ to
be open if $\mathcal{O} \cap M_i$ is open for all $i \in I$.

  (i) Show that this defines a topology $\mathcal{M}$ on M, called the *disconnected sum* topol-
       ogy.
 (ii) Show that the subsets $M_i, M_j \subseteq M$ are in different connected components of
       M for $i \neq j$.
(iii) Show that the canonical inclusion map $\iota_i \colon M_i \longrightarrow M$ is a continuous injection.
       Show that the subspace topology coincides with the original topology $\mathcal{M}_i$, i.e.
       ι_i is an embedding.

(iv) Show that the disconnected sum topology is the final topology on M with respect to the maps ι_i for $i \in I$.

Exercise 3.4.6 (The sphere $\mathbb{S}^n$ as topological manifold) Consider the n-dimensional sphere $\mathbb{S}^n$ with Euclidean radius 1 around the origin in $\mathbb{R}^{n+1}$, see (2.1.8). As usual, we endow $\mathbb{S}^n$ with the subspace topology. Define the north and south pole by $N = (0, \ldots, 0, 1)$ and by $S = (0, \ldots, 0, -1)$, respectively. Moreover, from N or S one defines the *stereographic projection* x_N and x_S onto the hyperplane $\mathbb{R}^n \subseteq \mathbb{R}^{n+1}$ which is determined by $x^{n+1} = 0$: a point $p \in \mathbb{S}^n \setminus \{N\}$ is mapped to the intersection $x_N(p) \in \mathbb{R}^n$ of the unique straight line through N and p with the hyperplane $\mathbb{R}^n$. Analogously, one defines x_S.

(i) First, consider the cases $n = 1$ and $n = 2$ and sketch the stereographic projections in this situation.
(ii) For general n compute $x_N(p)$ and $x_S(p)$ explicitly.
(iii) Show that x_N is a homeomorphism from $\mathbb{S}^n \setminus \{N\}$ onto its image $\mathbb{R}^n$ by finding an explicit formula for the inverse map.
(iv) Conclude that $\mathbb{S}^n$ is a topological manifold of dimension n.
(v) Find the domain of definition for the change of coordinates from the north pole chart to the south pole chart. Show that this is not only a homeomorphism but even a diffeomorphism, i.e. an infinitely often differentiable map.

Exercise 3.4.7 (The sphere $\mathbb{S}^n$ as a quotient) Consider again the sphere $\mathbb{S}^n \subseteq \mathbb{R}^{n+1}$ with its canonical subspaces topology.

(i) Consider the multiplicative group $\mathbb{R}^+$ and define

$$\Phi \colon \mathbb{R}^+ \times \mathbb{R}^{n+1} \longrightarrow \mathbb{R}^{n+1} \tag{3.4.5}$$

by $\Phi(\lambda, x) = \lambda x$ for $\lambda \in \mathbb{R}^+$ and $x \in \mathbb{R}^{n+1}$. Show that Φ defines an action by homeomorphisms.
(ii) Show that $0 \in \mathbb{R}^{n+1}$ is a fixed point of the action Φ. Conclude that Φ restricts to an action of $\mathbb{R}^+$ on $\mathbb{R}^{n+1} \setminus \{0\}$.
(iii) Describe the orbit $\mathbb{R}^+ \cdot x$ for a vector $x \in \mathbb{R}^{n+1} \setminus \{0\}$.
(iv) Denote the orbit space of the action Φ on $\mathbb{R}^{n+1} \setminus \{0\}$ by $M = \left(\mathbb{R}^{n+1} \setminus \{0\}\right) / \mathbb{R}^+$ and endow it with the quotient topology. Show that

$$M \ni [x] \longmapsto \frac{x}{\|x\|} \in \mathbb{S}^n \tag{3.4.6}$$

defines a continuous bijection to the sphere $\mathbb{S}^n$. Determine the inverse map and show explicitly that this is continuous as well. This way, the sphere $\mathbb{S}^n$ can be identified with the orbit space $\left(\mathbb{R}^{n+1} \setminus \{0\}\right) / \mathbb{R}^+$.
(v) Consider now the whole orbit space $\mathbb{R}^{n+1} / \mathbb{R}^+$: show that the orbit of 0 can not be separated from any other orbit in the quotient topology. This gives an example where the quotient topology is not Hausdorff anymore, even though the original space had this separation property.

Exercise 3.4.8 (Properties of orbit spaces) Let $(M, \mathcal{M})$ be a topological space and let G be a group acting on M via homeomorphisms Φ_g. We denote the orbit space and the corresponding quotient map by $\pi \colon M \longrightarrow M/G$.

(i) Let $U \subseteq M$ be a subset. Show that

$$\pi^{-1}(\pi(U)) = \bigcup_{g \in G} \Phi_g(U). \tag{3.4.7}$$

(ii) Show that π is an open map.

(iii) Let $\mathcal{B}$ be a basis of the topology $\mathcal{M}$ of M. Show that the images of subsets from $\mathcal{B}$ in M/G from a basis of the quotient topology. Conclude that M/G is second countable if M was second countable.

(iv) Assume that G is finite and assume that M is Hausdorff. Prove that M/G is Hausdorff again.

Hint: This can be done either by a tedious careful construction of the separating subsets in the quotient relying on (i), or via the following trick: endow the group with the discrete topology, then the extended action map

$$\overline{\Phi} \colon G \times M \ni (g, p) \mapsto (\Phi_g(p), p) \in M \times M$$

is continuous. Let $A \subseteq M$ be closed. Show that the subset $\overline{\Phi}(\{g\} \times A)$ is homeomorphic to the image of the diagonal $\Delta(A) \subseteq M \times M$. Use that G is finite to conclude that $\overline{\Phi}$ is a closed map. Then show that $\overline{\Phi}(G \times M)$ is precisely the orbit relation and apply Proposition 3.2.6. This last part has an enormous generalization in the theory of group actions of topological groups and their proper actions. However, a detailed discussion would lead us much to far.

Exercise 3.4.9 (Cartesian products and maps) Let I be a non-empty index set and let $(M_i, \mathcal{M}_i)$ and $(N_i, \mathcal{N}_i)$ be topological spaces. Moreover, let $f_i \colon M_i \longrightarrow N_i$ be maps. Finally, endow $M = \prod_{i \in I} M_i$ and $N = \prod_{i \in I} N_i$ with the product topology.

(i) Show that the product map

$$f = \prod_{i \in I} f_i \colon \prod_{i \in I} M_i \longrightarrow \prod_{i \in I} N_i \tag{3.4.8}$$

is continuous iff all the maps f_i are continuous.

Hint: Use continuous sections $s_i \colon M_i \longrightarrow M$ as in the proof of Proposition 3.1.2 for every $i \in I$.

(ii) Show that the map f from (3.4.8) is a homeomorphism iff each f_i is a homeomorphism.

Exercise 3.4.10 (Products of topological manifolds) Let I be a finite index set.

(i) Show that for topological manifolds M_i their product $M = \prod_{i \in I} M_i$ is again a topological manifold. What is the dimension of M in terms of the dimensions of the M_i?

(ii) Show that the n-dimensional torus $\mathbb{T}^n \subseteq \mathbb{C}^n$ with its subspace topology is a topological manifold of dimension n.

Hint: Write $\mathbb{T}^n$ as a suitable Cartesian product and use part (i).

Exercise 3.4.11 (The projective space $\mathbb{RP}^n$) Consider for $n \in \mathbb{N}$ the space $\mathbb{R}^{n+1} \setminus \{0\}$ with the relation $\sim$ defined by $x \sim y$ iff there is a $\lambda \in \mathbb{R} \setminus \{0\}$ with $x = \lambda y$.

(i) Show that $\sim$ defines an equivalence relation and prove that the equivalence classes can be identified with the real lines in $\mathbb{R}^{n+1} \setminus \{0\}$.

(ii) Define the real projective space as the quotient $\mathbb{RP}^n = \mathbb{R}^{n+1} \setminus \{0\} / \sim$ and endow it with the quotient topology. Show that one can view $\mathbb{RP}^n$ also as quotient with respect to an appropriate group action by homeomorphisms of the multiplicative group $\mathbb{R} \setminus \{0\}$ on $\mathbb{R}^{n+1} \setminus \{0\}$.

(iii) Prove that $\mathbb{RP}^n$ is second countable.

(iv) Prove that $\mathbb{RP}^n$ is Hausdorff.

(v) Let $x = (x^0, \ldots, x^n)$ denote the canonical linear coordinates on $\mathbb{R}^{n+1} \setminus \{0\}$ and consider

$$U_k = \left\{ [x] \in \mathbb{RP}^n \;\middle|\; x^k \neq 0 \right\} \subseteq \mathbb{RP}^n. \tag{3.4.9}$$

Show that U_k is open for all $k = 0, \ldots, n$ and $\bigcup_{k=0}^n U_k = \mathbb{RP}^n$.

(vi) Define $y_k : U_k \longrightarrow \mathbb{R}^n$ by

$$y_k([x]) = \frac{1}{x^k} \left(x^0, \ldots, \overset{k}{\wedge}, \ldots, x^n \right) \tag{3.4.10}$$

for $[x] \in U_k$. Show that this is a homeomorphism from U_k to $\mathbb{R}^n$.

This way, $\mathbb{RP}^n$ becomes a topological manifold of dimension n. It is very illustrative to compute the change of coordinates when passing from the local chart (U_k, y_k) to the local chart (U_ℓ, y_ℓ) for $k \neq \ell$.

Exercise 3.4.12 (The projective space $\mathbb{CP}^n$) Consider $\mathbb{C}^{n+1} \setminus \{0\}$ with the relation that $z \sim w$ iff there is a $\lambda \in \mathbb{C} \setminus \{0\}$ with $z = \lambda w$. Formulate and prove the analogous statements as in Exercise 3.4.11 in order to show that the complex projective space $\mathbb{CP}^n = \mathbb{C}^{n+1} \setminus \{0\} / \sim$ is a topological manifold of (real) dimension $2n$. Construct the local charts (U_k, y_k) having their images in $\mathbb{C}^n$ and compute the corresponding change of coordinates.

Chapter 4
Convergence in Topological Spaces

In this chapter we will consider sequences in topological spaces and their convergence. For metric spaces, sequences will be the appropriate tool to study all phenomena of convergence and continuity. However, in a general topological space we will meet situations where sequences are simply not sufficient to investigate all questions of convergence and the relations to closures, continuity, and, later, the relations to compactness. One needs more general "sequences" than just those indexed by $\mathbb{N}$. This leads to the notion of *nets* (or: Moore-Smith sequences). Alternatively, convergence can be formulated using the concept of *filters*. We will define and compare these notions of convergence and discuss first properties of ultrafilters.

4.1 Convergence of Nets

Let $(M, \mathcal{M})$ be a topological space and let $(p_n)_{n \in \mathbb{N}}$ be a sequence of points $p_n \in M$ in this space. The basic idea of convergence is that p_n converges to $p \in M$ if for large n all the points p_n are close to p. Using the notion of neighbourhoods this is now easy to formulate more precisely: for all neighbourhoods $U \in \mathfrak{U}(p)$ there is an $N \in \mathbb{N}$ with $p_n \in U$ for all $n \geq N$. While this is a straightforward generalization of the usual concept of convergence of sequences, it will not be enough to consider sequences: we have to pass to nets.

Definition 4.1.1 *(Directed set)* Let I be a non-empty set and let $\preccurlyeq$ be a relation on I such that

(i) for all $i \in I$ one has $i \preccurlyeq i$,
(ii) for all $i, j, k \in I$ one has that $i \preccurlyeq j$ and $j \preccurlyeq k$ implies $i \preccurlyeq k$,
then $(I, \preccurlyeq)$ is called pre-ordered set. If in addition one has
(iii) for all $i, j \in I$ there exists a $k \in I$ with $i \preccurlyeq k$ and $j \preccurlyeq k$,
then $(I, \preccurlyeq)$ is called a directed set.

© Springer International Publishing Switzerland 2014

S. Waldmann, *Topology*, DOI 10.1007/978-3-319-09680-3_4

It is a good intuition to think of the relation $i \preccurlyeq j$ as "i is earlier than j" in the directed set $(I, \preccurlyeq)$. As already for the notion of finer topologies, we use $\preccurlyeq$ in the sense of "earlier or equal" but not in the sense of "strictly earlier". Note that we do not require that for all pairs (i, j) we can actually compare i and j, i.e. i and j might not be in relation with respect to $\preccurlyeq$. Moreover, one should be aware that there are many other notions of orders available. In particular, a *partial order* satisfies the properties (i) and (ii) together with the property

(iv) $i \preccurlyeq j$ and $j \preccurlyeq i$ implies $i = j$.

Note that for a directed set we do not require this additional property.

Example 4.1.2 (Directed sets)

(i) The set of natural numbers $\mathbb{N}$ is directed with respect to $\preccurlyeq \, = \, \leq$. Here $\leq$ is of course even a partial order.

(ii) Analogously, $\mathbb{Z}$, $\mathbb{Q}$, and $\mathbb{R}$ are directed (and partially ordered) with respect to $\preccurlyeq \, = \, \leq$.

(iii) Let M be a set and 2^M its power set. Then $\preccurlyeq \, = \, \subseteq$ defines a directed (and partially ordered) set $(2^M, \preccurlyeq)$. Indeed, it is clear that $\preccurlyeq$ satisfies the additional requirement of a partial order. Moreover, for $U, V \subseteq M$ we have $U \subseteq U \cup V$ and $V \subseteq U \cup V$, showing that $(2^M, \preccurlyeq)$ is directed.

(iv) If $(I, \preccurlyeq)$ and $(J, \preccurlyeq)$ are directed sets then $I \times J$ becomes directed by $(i, j) \preccurlyeq (i', j')$ if $i \preccurlyeq i'$ and $j \preccurlyeq j'$. We call this the *canonical direction* on the Cartesian product $I \times J$.

(v) More generally, if $(I, \preccurlyeq)$ is directed and X is an arbitrary non-empty set then $I \times X$ becomes directed by $(i, x) \preccurlyeq (i', x')$ if $i \preccurlyeq i'$. Here the presence of X is simply ignored. Note that if $(I, \preccurlyeq)$ was also partially ordered, this will typically no longer be true for $I \times X$.

(vi) Let $(M, \mathcal{M})$ be a topological space and let $p \in M$. Then the neighbourhoods $\mathfrak{U}(p)$ of p are directed by $U \preccurlyeq V$ if $V \subseteq U$ for $U, V \in \mathfrak{U}(p)$. Indeed, for $U, V \in \mathfrak{U}(p)$ also $U \cap V \subseteq \mathfrak{U}(p)$ and $U \preccurlyeq U \cap V$ as well as $V \preccurlyeq U \cap V$. This directed set will play a crucial role. Again, the neighbourhoods are even partially ordered. A slight variation shows that also a basis of neighbourhoods is a directed set with respect to the inherited $\preccurlyeq$. For these directed set one should think of the direction $U \preccurlyeq V$ as V is *closer* to p as U.

We come now to the definition of a net.

Definition 4.1.3 *(Net)* Let M be a set and let $(I, \preccurlyeq)$ be a directed set. Then a map

$$I \ni i \mapsto p_i \in M \tag{4.1.1}$$

is called a net in M indexed by I. We simply write $(p_i)_{i \in I}$ for such a map.

We will also need the notion of a subnet. This is somewhat more involved than the notion of a subsequence. If $(I, \preccurlyeq)$ and $(J, \preccurlyeq)$ are directed sets then a map

$$\Phi : I \longrightarrow J \tag{4.1.2}$$

is called *cofinal* if for every $j \in J$ there is an index $i \in I$ such that $i \preccurlyeq i'$ implies $j \preccurlyeq \Phi(i')$. The idea is that late elements in I become late elements in J. Using this, a *subnet* of a net $(p_j)_{j \in J}$ in M is a net of the form $(p_{\Phi(i)})_{i \in I}$ with a cofinal map $\Phi: I \longrightarrow J$. Sometimes it is also required that Φ is *monotonic*, i.e. $i \preccurlyeq i'$ implies $\Phi(i) \preccurlyeq \Phi(i')$. However, even though this will be the case in all relevant examples, we do not follow this convention for the definition of a subnet but just ask for a cofinal map Φ.

Remark 4.1.4 Clearly, a sequence is a particular net where the index set is just $\mathbb{N}$. Moreover, a subsequence of a sequence is a subnet in the above sense. However, a sequence can have subnets which are not subsequences. In particular, a cofinal map needs not to be injective at all, allowing for much larger index sets I for a subnet of a sequence than $\mathbb{N}$.

We can now define the convergence of nets analogously to the convergence of sequences:

Definition 4.1.5 *(Convergence of nets)* Let $(M, \mathcal{M})$ be a topological space and let $(p_i)_{i \in I}$ be a net in M. Then $(p_i)_{i \in I}$ converges to $p \in M$ if for every $U \in \mathfrak{U}(p)$ there is an index $i \in I$ with

$$p_j \in U \quad \text{for all} \quad i \preccurlyeq j. \tag{4.1.3}$$

The set of points to which $(p_i)_{i \in I}$ converges is called the limit set of the net and will be denoted by $\lim_{i \in I} p_i$. We write $p \in \lim_{i \in I} p_i$ for a limit point p with (4.1.3). We also write $p_i \longrightarrow p$ if $p \in \lim_{i \in I} P_i$.

In general, limits will not be unique, see also Exercise 4.4.2. The uniqueness of limits is intimately linked to the Hausdorff property:

Proposition 4.1.6 *Let $(M, \mathcal{M})$ be a topological space. Then $(M, \mathcal{M})$ is Hausdorff iff every net in M has at most one limit point.*

Proof Suppose $(M, \mathcal{M})$ is Hausdorff and let $(p_i)_{i \in I}$ be a convergent net with $p, q \in \lim_{i \in I} p_i$ but $p \neq q$. Then there are open neighbourhoods $U \in \mathfrak{U}(p)$ and $V \in \mathfrak{U}(q)$ with $U \cap V = \emptyset$. From convergence to p we get $i_1 \in I$ with $p_j \in U$ for $j \succcurlyeq i_1$ while from convergence to q we get $i_2 \in I$ with $p_j \in V$ for $j \succcurlyeq i_2$. Since I is directed we find a $i_3 \in I$ with $i_3 \succcurlyeq i_1$ and $i_3 \succcurlyeq i_2$. Thus for this i_3 we have $p_{i_3} \in U$ and $p_{i_3} \in V$, a contradiction. Thus $(p_i)_{i \in I}$ can have at most one limit. Conversely, assume $(M, \mathcal{M})$ is not Hausdorff and let $p, q \in M$ be distinct points which can not be separated. Then for all $U \in \mathfrak{U}(p)$ and $V \in \mathfrak{U}(q)$ we have $U \cap V \neq \emptyset$. From Example 4.1.2, (vi), we know that $\mathfrak{U}(p)$ and $\mathfrak{U}(q)$ are both directed via $\preccurlyeq = \subseteq$. From Example 4.1.2, (iv), we know that also $\mathfrak{U}(p) \times \mathfrak{U}(q)$ is directed. Since $U \cap V \neq \emptyset$ we find a point $p_{(U,V)} \in U \cap V$ for each $U \in \mathfrak{U}(p)$ and $V \in \mathfrak{U}(q)$ defining thereby a net $(p_{(U,V)})_{(U,V) \in \mathfrak{U}(p) \times \mathfrak{U}(q)}$. Now if $U \in \mathfrak{U}(p)$ is given, then for all later $U' \succcurlyeq U$ we have $U' \subseteq U$ and thus $p_{(U',V)} \in U' \subseteq U$ for all (U', V). This shows $p_{U,V} \longrightarrow p$. By symmetry we also get $p_{U,V} \longrightarrow q$, contradicting $p \neq q$. $\qquad\square$

The next proposition shows that we can obtain the closure of a subset by approaching the boundary points by means of net convergence:

Proposition 4.1.7 *Let* $(M, \mathcal{M})$ *be a topological space and let* $A \subseteq M$. *Then* $p \in A^{\mathrm{cl}}$ *iff there exists a net* $(p_i)_{i \in I}$ *with* $p_i \in A$ *converging to* p.

Proof Let $p \in A^{\mathrm{cl}}$. Then for every $U \in \mathfrak{U}(p)$ we find a point $p_U \in U \cap A$. This defines a net $(p_U)_{U \in \mathfrak{U}(p)}$ converging to p as wanted. Conversely, let $p_i \in A$ for $i \in I$ and $p_i \longrightarrow p$. Then for every $U \in \mathfrak{U}(p)$ there is an index $i \in I$ with $j \succcurlyeq i$ implies $p_j \in U$. Thus in particular $U \cap A \neq \emptyset$, showing $p \in A^{\mathrm{cl}}$. $\qquad\square$

In general, $\mathfrak{U}(p)$ may be too large to allow this statement also for sequences instead of nets. This motivates the following definition:

Definition 4.1.8 *(Sequential closure)* Let $(M, \mathcal{M})$ be a topological space and $A \subseteq M$. Then the sequential closure $A^{\mathrm{scl}} \subseteq M$ is the set of points in M which arise as limits of convergent sequences $(p_n)_{n \in \mathbb{N}}$ with $p_n \in A$.

Proposition 4.1.9 *Let* $(M, \mathcal{M})$ *be a topological space and* $A \subseteq M$.

(i) We have $A \subseteq A^{\mathrm{scl}} \subseteq A^{\mathrm{cl}}$.
(ii) If $(M, \mathcal{M})$ *is first countable then* $A^{\mathrm{scl}} = A^{\mathrm{cl}}$.

Proof The first part is clear by Proposition 4.1.7. For the second statement we note that the same argument in the proof of Proposition 4.1.7 allows to construct a sequence by considering only a (countable) basis of $\mathfrak{U}(p)$ instead of $\mathfrak{U}(p)$ as index set. The sequence then converges to the point p in the closure, see also Exercise 4.4.3. $\qquad\square$

The importance of nets is also illustrated by the following Proposition:

Proposition 4.1.10 *Let* $f : (M, \mathcal{M}) \longrightarrow (N, \mathcal{N})$ *be a map between topological spaces. Then the following statements are equivalent:*

(i) The map f *is continuous.*
(ii) For every convergent net $(p_i)_{i \in I}$ *in* M *also the net* $(f(p_i))_{i \in I}$ *is convergent in* N *and*

$$\lim_{i \in I} f(p_i) = f\left(\lim_{i \in I} p_i\right). \tag{4.1.4}$$

Proof We prove a slightly more specific statement: first assume that f is continuous at $p \in M$ and let $p_i \longrightarrow p$. Let $V \in \mathfrak{U}(f(p))$ be a neighbourhood of $f(p)$ and let $U \in \mathfrak{U}(p)$ be a neighbourhood of p with $f(U) \subseteq V$, by continuity at p. Then we have an $i \in I$ with $p_j \in U$ for all $j \succcurlyeq i$ by convergence. Thus also $f(p_j) \in V$ for all those $j \succcurlyeq i$ showing $f(p_i) \longrightarrow f(p)$. Conversely, assume that for every net $(p_i)_{i \in I}$ with $p_i \longrightarrow p$ we have $f(p_i) \longrightarrow f(p)$. Assume that f is not continuous at p. Then there is a neighbourhood $V \in \mathfrak{U}(f(p))$ for which $f^{-1}(V)$ is not a neighbourhood of p. Thus for every $U \in \mathfrak{U}(p)$ there is a point $p_U \in U$ with $f(p_U) \notin V$. Then the net $(p_U)_{U \in \mathfrak{U}(p)}$ converges to p but $(f(p_U))_{U \in \mathfrak{U}(p)}$ does not converge to $f(p)$, a contradiction. $\qquad\square$

Rephrasing the proposition, we can say that continuity (at a point) is equivalent to *net continuity* (at that point). Again, sequences will not be sufficient, so *sequential continuity* does no imply continuity in general. As for the sequential closure, we need the first countability axiom for this:

Proposition 4.1.11 *Let $f: (M, \mathcal{M}) \longrightarrow (N, \mathcal{N})$ be a map between topological spaces and assume that $(M, \mathcal{M})$ is first countable. Then f is continuous (at $p \in M$) iff f is sequentially continuous (at $p \in M$).*

Proof By Proposition 4.1.10 continuity implies sequential continuity, as sequences are particular nets. For the converse direction, we note that it suffices to consider a countable basis of neighbourhoods of p in the proof of Proposition 4.1.10, and hence a sequence.   □

We conclude this subsection with a useful statement on the convergence of subnets:

Proposition 4.1.12 *Let $(M, \mathcal{M})$ be a topological space and let $(p_j)_{j \in J}$ be a net in M. Then $(p_j)_{j \in J}$ converges to p iff every subnet $(p_{\Phi(i)})_{i \in I}$ of $(p_j)_{j \in J}$ converges to p.*

Proof Assume that $p_j \longrightarrow p$ and let $\Phi : I \longrightarrow J$ be a cofinal map for some other directed set I. Let $U \in \mathfrak{U}(p)$ be a neighbourhood of p. Then there is a $j \in J$ with $p_{j'} \in U$ for all $j' \succcurlyeq j$ by convergence. Since Φ is cofinal, we get an index $i \in I$ with $i' \succcurlyeq i$ implies $\Phi(i') \succcurlyeq j$. Thus $p_{\Phi(i')} \in U$ and we have convergence $p_{\Phi(i)} \longrightarrow p$. The other direction is trivial as $(p_j)_{j \in J}$ is a subnet of itself.   □

4.2 Nets and Filters

We have already seen that the system of neighbourhoods $\mathfrak{U}(p)$ of a point $p \in M$ or even a basis of neighbourhoods of p can be used effectively as indexing set for a net. This is the starting point to establish yet another concept of convergence, the convergence of filters.

Definition 4.2.1 *(Filter)* Let M be a set and $\mathfrak{F} \subseteq 2^M$ a collection of subsets of M.

(i) $\mathfrak{F}$ is called a filter if

- $\emptyset \notin \mathfrak{F}$,
- $A, B \in \mathfrak{F}$ implies $A \cap B \in \mathfrak{F}$,
- $A \in \mathfrak{F}$ and $A \subseteq B$ implies $B \in \mathfrak{F}$.

(ii) A filter $\mathfrak{F}$ is finer than a filter $\mathfrak{F}'$ if $\mathfrak{F}' \subseteq \mathfrak{F}$.
(iii) A subset $\mathfrak{B} \subseteq \mathfrak{F}$ of a filter is called filter basis of $\mathfrak{F}$ if for all $A \in \mathfrak{F}$ there exists a $B \in \mathfrak{B}$ with $B \subseteq A$.

(iv) A filter $\mathfrak{F}$ is called free if

$$\bigcap_{A \in \mathfrak{F}} A = \emptyset, \qquad\qquad (4.2.1)$$

and fixed otherwise.

(v) A filter $\mathfrak{F}$ is called an ultrafilter if there is no finer filter than $\mathfrak{F}$ beside $\mathfrak{F}$ itself.

As for topologies we note that "finer" is used in the sense of "finer or equal" throughout this text.

Example 4.2.2 (Filters)

(i) Let $A \subseteq M$ be a non-empty subset and define

$$\mathfrak{F} = \{B \subseteq M \mid A \subseteq B\}. \qquad\qquad (4.2.2)$$

Then $\mathfrak{F}$ defines a filter which is not free as A is contained in all sets belonging to $\mathfrak{F}$. Such a filter is also called the *principal filter* of the subset A.

(ii) Let $(M, \mathcal{M})$ be a topological space and $p \in M$. Then $\mathfrak{U}(p)$ is a filter, the *neighbourhood filter* of p. Again, this filter is not free as $p \in \bigcap_{U \in \mathfrak{U}(p)} U$ by the very definition of neighbourhoods.

(iii) Let M be an infinite set and define

$$\mathfrak{F} = \left\{A \in M \mid M \setminus A \text{ is finite}\right\}. \qquad\qquad (4.2.3)$$

This is a filter which is free.

(iv) If $f : M \longrightarrow N$ is a map and $\mathfrak{F}$ a filter on M then

$$f_*\mathfrak{F} = \left\{B \subseteq N \mid \text{there is a } A \in \mathfrak{F} \text{ with } f(A) \subseteq B\right\} \qquad\qquad (4.2.4)$$

is a filter on N, the *push-forward* or *image* of $\mathfrak{F}$ by f, see also Exercise 4.4.9.

(v) Let M be a set and $p \in M$. Then repeating the construction of (i) for the subset $\{p\}$ gives an ultrafilter which is fixed. If on the other hand, $A \subseteq M$ contains more than one point then the filter (4.2.2) is not an ultrafilter.

Nets and filters are closely related. In fact, we can construct a net from every filter and vice versa:

Proposition 4.2.3 *Let M be a set.*

(i) *If $\mathfrak{F}$ is a filter on M then the set $I_{\mathfrak{F}}$ of pairs (A, p) with $A \in \mathfrak{F}$ and $p \in A$ is directed via $(A, p) \preccurlyeq (B, q)$ if $A \supseteq B$ and we have a net*

$$I_{\mathfrak{F}} \ni (A, p) \mapsto p \in M, \qquad\qquad (4.2.5)$$

called the net associated to the filter $\mathfrak{F}$.

(ii) If $(p_i)_{i \in I}$ is a net in M then

$$\mathfrak{F} = \{A \subseteq M \mid \text{there is an } i \in I \text{ with } p_j \in A \text{ for all } j \succcurlyeq i\} \qquad (4.2.6)$$

is a filter on M, called the filter associated to the net $(p_i)_{i \in I}$. The subsets

$$A_i = \{p_j \mid j \succcurlyeq i\} \qquad (4.2.7)$$

form a basis of $\mathfrak{F}$.

(iii) Let $(p_i)_{i \in I}$ be a net with associated filter $\mathfrak{F}$. Then $(p_i)_{i \in I}$ is a subnet of the net associated to the filter $\mathfrak{F}$.

(iv) Let $\mathfrak{F}$ be a filter and let $\mathfrak{G}$ be the filter associated to the net associated to $\mathfrak{F}$. Then $\mathfrak{F} = \mathfrak{G}$.

Proof The fact that $I_{\mathfrak{F}}$ is directed follows essentially the line of argument in Example 4.1.2, (vi), since again $A \cap B \in \mathfrak{F}$ for $A, B \in \mathfrak{F}$. Then the first part is clear. For the second we note that all subsets $A \in \mathfrak{F}$ are non-empty. If $A, B \in \mathfrak{F}$ with $i, i' \in I$ such that $p_j \in A$ and $p_{j'} \in B$ for all $j \succcurlyeq i$ and $j' \succcurlyeq i'$ then take $i'' \succcurlyeq i, i'$ to conclude that $j'' \succcurlyeq i''$ implies $p_{j''} \in A \cap B$. Here we use that I is directed. Thus $A \cap B \in \mathfrak{F}$. The remaining third property of a filter is clear, showing the second statement. For the third part, define the map

$$I \ni i \longmapsto (A_i, p_i) \in I_{\mathfrak{F}}, \qquad (*)$$

where A_i is the subset as in (4.2.7). This is a cofinal map. Indeed, for any $(A, p) \in I_{\mathfrak{F}}$ there is an index $i \in I$ with $p_j \in A$ for all $j \succcurlyeq i$ and hence for this i we have $A_i \subseteq A$ meaning $(A_i, p_i) \succcurlyeq (A, p)$. Note that $(*)$ is also monotonic. This way, we can realize the net $(p_i)_{i \in I}$ as subnet of the net associated to the filter $\mathfrak{F}$ as in (4.2.6). For the last part, we consider $A \in \mathfrak{G}$. This means that there is an index $(B, p) \in I_{\mathfrak{F}}$ such that for all later indices $(B', p') \succcurlyeq (B, p)$ we have $p' \in A$. Now $(B, p) \in I_{\mathfrak{F}}$ means $p \in B \subseteq \mathfrak{F}$ showing $B \subseteq A$. Thus $A \in \mathfrak{F}$ again and $\mathfrak{G} \subseteq \mathfrak{F}$ follows. Conversely, if $A \in \mathfrak{F}$ then take $B = A$ and p some point in A. Then for all $(B', p') \succcurlyeq (A, p)$ we have $p' \in B' \subseteq A$ and thus $p' \in A$. This shows $A \in \mathfrak{G}$, completing the proof. $\square$

Thus we can always recover a filter from its associated net but not a net from its associated filter: here we recover only a "larger" net. Thus there may be different nets having the same associated filter in which case they are all subnets of the associated net of this filter. In view of Proposition 4.1.12 this will be only a minor problem when it comes to convergence.

Lemma 4.2.4 *Let $f \colon M \longrightarrow N$ be a map.*

(i) The push-forward of the filter associated to a net $(p_i)_{i \in I}$ in M and the filter associated to the net $(f(p_i))_{i \in I}$ coincide.

(ii) The image of the net associated to a filter $\mathfrak{F}$ in M is a subnet of the net associated to the push-forward filter $f_* \mathfrak{F}$.

Proof For the first part, let $B \subseteq N$ be in the image filter associated to $(p_i)_{i \in I}$. This means there is an $A \subseteq M$ with $f(A) \subseteq B$ and an $i \in I$ such that $p_j \in A$ for $j \succcurlyeq i$. Hence $f(p_j) \in B$ for $j \succcurlyeq i$ and thus B belongs to the filter associated to $(f(p_i))_{i \in I}$. Conversely, let $B \subseteq N$ be in the filter associated to $(f(p_i))_{i \in I}$. Then there is an $i \in I$ with $f(p_j) \in B$ for $j \succcurlyeq i$. But this shows $f(A_i) \subseteq B$, proving the first part. For the second part, let $(A, p) \in I_{\mathfrak{F}}$ with the corresponding net $(A, p) \mapsto p$. Then the image net is $(f(p))_{(A,p) \in I_{\mathfrak{F}}}$. Defining

$$\Phi : I_{\mathfrak{F}} \ni (A, p) \ \mapsto \ (f(A), f(p)) \in I_{f_*\mathfrak{F}}$$

gives a map between the two index sets such that $(f(p_{(A,p)}))$ is given by $(f(p))_{\Phi(A,p)}$ viewed as subnet of the net associated to $f_*\mathfrak{F}$. Note that Φ is indeed cofinal.   □

We can now define convergence of a filter in a topological space such that filter convergence coincides with the convergence of its associated net. In view of Proposition 4.1.12 this matches with the correspondence of filters and nets. Traditionally, one proceeds in a slightly different way:

Definition 4.2.5 *(Convergence of filters)* Let $(M, \mathcal{M})$ be a topological space and let $\mathfrak{F}$ be a filter on M. Then $\mathfrak{F}$ converges to $p \in M$ if $\mathfrak{F}$ is finer than the neighbourhood filter $\mathfrak{U}(p)$ of p.

Proposition 4.2.6 *Let $(M, \mathcal{M})$ be a topological space and $p \in M$.*

  (i)   The neighbourhood filter $\mathfrak{U}(p)$ converges to p.
 *(ii)   If a filter $\mathfrak{F}$ converges to p and a filter $\mathfrak{F}'$ is finer than $\mathfrak{F}$, then $\mathfrak{F}'$ converges to
        p, too.*
(iii)   A filter converges to p iff its associated net converges to p.
 (iv)   A net converges to p iff its associated filter converges to p.

Proof The first part is clear and so is the second. For the third, let $\mathfrak{F}$ converge to p. Then for every $U \in \mathfrak{U}(p)$ we have $U \in \mathfrak{F}$. Thus for all indices $(A, q) \in I_{\mathfrak{F}}$ with $(A, q) \succcurlyeq (U, p)$ we have $q \in U$. Thus the net associated to $\mathfrak{F}$ converges to p. Conversely, assume that the net $(q_{(A,q)})$ with $(A, q) \in I_{\mathfrak{F}}$ converges to p. Then for $U \in \mathfrak{U}(p)$ we have an index (A, q) such that for all $(A', q') \succcurlyeq (A, q)$ we have $q' \in U$. Hence in particular $A' \subseteq U$ and also $A \subseteq U$. But this means $U \in \mathfrak{F}$ and thus $\mathfrak{F}$ is finer than $\mathfrak{U}(p)$, showing the third part. Now suppose that the net $(p_i)_{i \in I}$ converges to p. Then for every neighbourhood $U \in \mathfrak{U}(p)$ we have an index i with $p_j \in U$ for $j \succcurlyeq i$. Hence the subset A_i from (4.2.7) is contained in U. This shows that the filter associated to $(p_i)_{i \in I}$ is finer than $\mathfrak{U}(p)$. For the converse, we can rely on the third part and Proposition 4.2.3 and Proposition 4.1.12.   □

The equivalence of net convergence and filter convergence gives several corollaries to the results from Sect. 4.1. We just list the results:

Corollary 4.2.7 *A topological space $(M, \mathcal{M})$ is Hausdorff iff every filter in M has at most one limit point.*

Corollary 4.2.8 *A map* $f : (M, \mathcal{M}) \longrightarrow (N, \mathcal{N})$ *between topological spaces is continuous at a point* $p \in M$ *iff every convergent filter* $\mathfrak{F}$ *converging to* p *has a convergent image filter* $f_* \mathfrak{F}$ *converging to* $f(p)$.

The last concept we need is that of a cluster point:

Definition 4.2.9 *(Cluster point)* Let $(M, \mathcal{M})$ be a topological space and $p \in M$. Then p is called a cluster point of a filter $\mathfrak{F}$ if for all $U \in \mathfrak{U}(p)$ and $A \in \mathfrak{F}$ one has $A \cap U \neq \emptyset$.

Clearly, any limit point p of a filter is also a cluster point as in this case even $\mathfrak{U}(p) \subseteq \mathfrak{F}$. Conversely, any cluster point can be obtained at least as a limit point of a finer filter:

Proposition 4.2.10 *Let* $(M, \mathcal{M})$ *be a topological space and* $p \in M$. *Then* p *is a cluster point of the filter* $\mathfrak{F}$ *iff there is a finer filter* $\mathfrak{G}$ *than* $\mathfrak{F}$ *which converges to* p.

Proof Assume that p is a cluster point of $\mathfrak{F}$. Then $U \cap A \neq \emptyset$ for all $U \in \mathfrak{U}(p)$ and $A \in \mathfrak{F}$. We take these sets as basis of a new filter $\mathfrak{G}$: indeed, this gives a well-defined filter as finite intersections of sets of the form $U \cap A$ are still non-empty. In fact they are again of this form. Thus taking all $B \subseteq M$ with $U \cap A \subseteq B$ for some $U \in \mathfrak{U}(p)$ and $A \in \mathfrak{F}$ defines a filter $\mathfrak{G}$, see also Exercise 4.4.8. This filter is finer than both filters $\mathfrak{U}(p)$ and $\mathfrak{F}$. Thus $\mathfrak{G} \longrightarrow p$. Conversely, let $\mathfrak{F} \subseteq \mathfrak{G}$ with $\mathfrak{G} \longrightarrow p$. For every $A \in \mathfrak{F}$ and $U \in \mathfrak{U}(p)$ we have $A, U \in \mathfrak{G}$ and hence $A \cap U \in \mathfrak{G}$, showing $A \cap U \neq \emptyset$. Hence p is a cluster point of $\mathfrak{F}$. □

Analogously, we can define cluster points of nets and get the result that p is a cluster point iff a subnet converges to p, see Exercise 4.4.6.

4.3 Ultrafilters

We have seen already some examples of ultrafilters in Example 4.2.2, (v). Using Zorn's Lemma one can show the existence of ultrafilters in the following proposition:

Proposition 4.3.1 *Let* $\mathfrak{F}$ *be a filter on a set* M. *Then there exists an ultrafilter containing* $\mathfrak{F}$.

Proof We consider the set Φ of all filters which contain $\mathfrak{F}$, i.e. which are finer than $\mathfrak{F}$. Since $\mathfrak{F} \in \Phi$ the set Φ is non-empty. We can partially order Φ by the inclusion relation "$\subseteq$" as usual. Now let $\{\mathfrak{F}_i\}_{i \in I}$ be a linearly ordered subset of Φ, i.e. filters with $\mathfrak{F}_i \subseteq \mathfrak{F}_j$ for $i \preccurlyeq j$ in the linearly ordered set I. Then we consider the union $\bigcup_{i \in I} \mathfrak{F}_i$, which is again a filter containing $\mathfrak{F}$. Indeed, let $A, B \in \bigcup_{i \in I} \mathfrak{F}_i$. Then there is one $j \in I$ with $A, B \in \mathfrak{F}_j$ as I is linearly ordered. Thus $A \cap B \in \mathfrak{F}_j \subseteq \bigcup_{i \in I} \mathfrak{F}_i$ since $\mathfrak{F}_j$ is a filter. Analogously, $\emptyset \notin \bigcup_{i \in I} \mathfrak{F}_i$ and $A \in \bigcup_{i \in I} \mathfrak{F}_i$ with $A \subseteq B$ implies $B \in \bigcup_{i \in I} \mathfrak{F}_i$. Thus every linearly ordered subset $\{\mathfrak{F}_i\}_{i \in I}$ of Φ has a supremum $\bigcup_{i \in I} \mathfrak{F}_i$. This allows now to apply Zorn's Lemma to show that there are maximal elements in Φ. It is clear that such a filter is indeed an ultrafilter finer than $\mathfrak{F}$. □

The next proposition gives some first properties and characterizations of ultrafilters:

Proposition 4.3.2 *Let $\mathfrak{F}$ be a filter on a set M.*

(i) The filter $\mathfrak{F}$ is an ultrafilter iff for every $A \subseteq M$ one has either $A \in \mathfrak{F}$ or $M \setminus A \in \mathfrak{F}$.

(ii) The filter $\mathfrak{F}$ is a fixed ultrafilter iff there is a point $p \in M$ with $\mathfrak{F} = \{A \subseteq M \mid p \in A\}$.

(iii) If $f : M \longrightarrow N$ is a map and $\mathfrak{F}$ an ultrafilter, then $f_* \mathfrak{F}$ is an ultrafilter on N.

Proof Suppose first that $\mathfrak{F}$ is an ultrafilter such that $M \setminus A$ is not in $\mathfrak{F}$. Now suppose that there is a $B \in \mathfrak{F}$ with $B \cap A = \emptyset$ then $B \subseteq M \setminus A$ showing $M \setminus A \in \mathfrak{F}$, a contradiction. Hence all elements of $\mathfrak{F}$ intersect non-trivially with A. But then we can define a new filter by taking $\{B \cap A \mid B \in \mathfrak{F}\}$ as basis. This filter contains all elements $B \in \mathfrak{F}$ and it contains A. Since it is finer than the ultrafilter $\mathfrak{F}$, it coincides with $\mathfrak{F}$ and $A \in \mathfrak{F}$ follows. Conversely, assume that for all $A \subseteq M$ either $A \in \mathfrak{F}$ or $M \setminus A \in \mathfrak{F}$. Assume that $\mathfrak{F} \subseteq \mathfrak{G}$ and assume that $\mathfrak{G}$ is a strictly finer filter. Then there is an element $A \in \mathfrak{G}$ with $A \notin \mathfrak{F}$. But then $M \setminus A \in \mathfrak{F} \subseteq \mathfrak{G}$. This contradicts the filter property as $A \cap (M \setminus A) = \emptyset$ can not be in $\mathfrak{G}$. It follows that $\mathfrak{F}$ is an ultrafilter. For the second part consider a fixed ultrafilter $\mathfrak{F}$, and let $B_0 = \bigcap_{B \in \mathfrak{F}} B$, which is non-empty by assumption. Thus there exists a $B \in \mathfrak{F}$ with $B_0 \cap B \neq \emptyset$. From the first part we conclude that $B_0 \in \mathfrak{F}$. Thus $\{B_0\}$ is a filter basis for $\mathfrak{F}$ and we conclude $\mathfrak{F} = \{B \subseteq M \mid B_0 \subseteq B\}$. From Example 4.2.2, (v), we know that $B_0 = \{p\}$ can contain at most one point. The converse is already contained in Example 4.2.2, (v). For the last part, let $B \subseteq N$. Then $f^{-1}(B) \subseteq M$ is either in $\mathfrak{F}$ or $M \setminus f^{-1}(B) = f^{-1}(N \setminus B)$ is in $\mathfrak{F}$. Suppose $f^{-1}(B) \in \mathfrak{F}$. Then $f(f^{-1}(B)) \subseteq B$ shows $B \in f_* \mathfrak{F}$. Conversely, suppose $f^{-1}(N \setminus B) \in \mathfrak{F}$. Then $f(f^{-1}(N \setminus B)) \subseteq N \setminus B$ shows $N \setminus B \in f_* \mathfrak{F}$. By the first part we conclude that $f_* \mathfrak{F}$ is an ultrafilter. □

As a last feature of ultrafilters we note that there is no difference between cluster points and limit points:

Proposition 4.3.3 *Let $(M, \mathcal{M})$ be a topological space and $\mathfrak{F}$ and ultrafilter on M. Then $p \in M$ is a cluster point of $\mathfrak{F}$ iff $\mathfrak{F}$ converges to p.*

Proof This is clear by Proposition 4.2.10. □

4.4 Exercises

Exercise 4.4.1 (Complete metric spaces) Let (M, d) be a complete metric space and let $A \subseteq M$ be a subset. Show that A is closed iff $(A, d|_A)$ is a complete metric space, too.

Exercise 4.4.2 (Convergence in the indiscrete space) Let M be a non-empty set equipped with the indiscrete topology. Show that any net in M converges to every point.

Exercise 4.4.3 (Net convergence and neighbourhood bases) Let $(M, \mathcal{M})$ be a topological space and let $p \in M$. Moreover, let $\mathfrak{B}(p) \subseteq \mathfrak{U}(p)$ be a basis of neighbourhoods of p.

(i) Prove that a net $(p_i)_{i \in I}$ converges to p iff for every $U \in \mathfrak{B}(p)$ there is an index $i \in I$ such that for all $j \succcurlyeq i$ one has $p_j \in U$.

(ii) Show that $\mathfrak{B}(p)$ is a directed set with the direction inherited from $\mathfrak{U}(p)$.

(iii) Let $A \subseteq M$ be a subset. Show that $p \in A^{\mathrm{cl}}$ iff there is a net $(p_U)_{U \in \mathfrak{B}(p)}$ indexed by the basis $\mathfrak{B}(p)$ with $p_U \longrightarrow p$.

Exercise 4.4.4 (Convergence in a Cartesian product) Let $\{(M_i, \mathcal{M}_i)\}_{i \in I}$ be non-empty topological spaces where I is an arbitrary non-empty index set.

(i) Consider nets $\left(p_j^{(i)}\right)_{j \in J}$ in M_i for all $i \in I$, where the index set J is the same for all $i \in I$. Consider the corresponding net $(p_j)_{j \in J}$ in M with i-th component being the j-th element given by $p_j^{(i)}$. Apparently, every net in M is of this form with $p_j^{(i)} = \mathrm{pr}_i(p_j)$. Show that $(p_j)_{j \in J}$ converges to $p \in M$ iff $\left(p_j^{(i)}\right)_{j \in J}$ converges to $p^{(i)} = \mathrm{pr}_i(p)$ for all $i \in I$.

(ii) Let $\mathfrak{F}$ be a filter on M and denote by $\mathfrak{F}_i = (\mathrm{pr}_i)_* \mathfrak{F}$ the corresponding image filters on each component M_i. Show that $\mathfrak{F}$ converges to $p \in M$ iff $\mathfrak{F}_i$ converges to $p_i = \mathrm{pr}_i(p)$ for all $i \in I$.

(iii) Let $A_i \subseteq M_i$ be subsets for every $i \in I$ and consider $A = \prod_{i \in I} A_i$ as subset of M. Show that

$$A^{\mathrm{cl}} = \prod_{i \in I} A_i^{\mathrm{cl}}. \qquad (4.4.1)$$

Hint: Use part (i) as well as Proposition 4.1.7.

Exercise 4.4.5 (Exchanging limits) Consider a T_3-space $(M, \mathcal{M})$ and two directed sets I and J. A direction $\preccurlyeq$ on the Cartesian product $I \times J$ is called *weaker* as the canonical direction $\preccurlyeq_{\mathrm{can}}$ from Example 4.1.2, (iv), if $(i, j) \preccurlyeq (i', j')$ follows from $(i, j) \preccurlyeq_{\mathrm{can}} (i', j')$. Let now such a weaker direction $\preccurlyeq$ on $I \times J$ be given. Consider a net $(p_{ij})_{(i,j) \in I \times J}$ in M. Assume that $(p_{ij})_{(i,j) \in I \times J}$ converges to $p \in M$. In addition, assume that for every fixed $i \in I$ the net $(p_{ij})_{j \in J}$ converges to a point $p_i \in M$. This defines a new net $(p_i)_{i \in I}$. Show that this net converges, too, and

$$\lim_{i \in I} p_i = p. \qquad (4.4.2)$$

Hint: For showing the convergence of $(p_i)_{i \in I}$ to p is suffices to consider *closed* neighbourhoods of p.

This statement can be viewed as an *exchanging of limits* once one considers the symmetric situation: suppose that also $p_j = \lim_{i \in I} p_{ij}$ exists for all $j \in J$. With this additional assumption one has

$$\lim_{i \in I} \lim_{j \in J} p_{ij} = \lim_{(i,j) \in I \times J} p_{ij} = \lim_{j \in J} \lim_{i \in I} p_{ij}. \tag{4.4.3}$$

Note that on $I \times J$ one can of course use the canonical direction $\preccurlyeq_{\mathrm{can}}$ as a particular case.

Exercise 4.4.6 (Cluster points of nets) Consider a topological space $(M, \mathcal{M})$ with a net $(p_i)_{i \in I}$. For a subset $A \subseteq M$ one says that the net $(p_i)_{i \in I}$ is *cofinal* in A, if for every index $i \in I$ there is a later index $j \succcurlyeq i$ with $p_j \in A$. A point $p \in M$ is called *cluster point* of the net $(p_i)_{i \in I}$ if the net is cofinal in every neighbourhood $U \in \mathfrak{U}(p)$ of p.

(i) Let $\mathfrak{F}$ be the filter associated to $(p_i)_{i \in I}$. Show that $p \in M$ is a cluster point of the net $(p_i)_{i \in I}$ iff p is a cluster point of $\mathfrak{F}$.

(ii) Conversely, let $\mathfrak{F}$ be a filter with associated net $(p_i)_{i \in I_{\mathfrak{F}}}$. Show that $p \in M$ is a cluster point of $\mathfrak{F}$ iff p is a cluster point of the net $(p_i)_{i \in I_{\mathfrak{F}}}$.

(iii) Show that a point $p \in M$ is a cluster point of a net $(p_i)_{i \in I}$ iff there is a subnet $(p_{\Phi(j)})_{j \in J}$ with a cofinal map $\Phi \colon J \longrightarrow I$ which is converging to p.

(iv) Let $(p_{\Phi(j)})_{j \in J}$ be a subnet of $(p_i)_{i \in I}$. Show that a cluster point of the subnet $(p_i)_{i \in I}$ is also a cluster point of the net $(p_i)_{i \in I}$. Is the converse also true?

Exercise 4.4.7 (The Riemann integral) Consider a bounded function $f \colon [a, b] \longrightarrow \mathbb{R}$ for $a < b$. A *partition* Z of $[a, b]$ with the length $n = |Z|$ is a finite set of points $a = t_0 < t_1 < \cdots < t_n = b$ between a and b. The set of all partitions of $[a, b]$ with length n will be denoted by $\mathfrak{Z}_n([a, b])$. Moreover, we set $\mathfrak{Z}([a, b]) = \bigcup_{n \in \mathbb{N}} \mathfrak{Z}_n([a, b])$. Now we consider the following set

$$I = \big\{ (Z, \xi) \mid Z \in \mathfrak{Z}([a, b]) \text{ and } \xi = (\xi_1, \ldots, \xi_n)$$
$$\text{with } \xi_i \in [t_{i-1}, t_i] \text{ for all } i = 1, \ldots, n = |Z| \big\}, \tag{4.4.4}$$

which will be ordered by $(Z, \xi) \succcurlyeq (Z', \xi')$ if $Z' \subseteq Z$.

(i) Show that I is directed via $\succcurlyeq$.

(ii) Define the net of Riemann sums of f by

$$I \ni (Z, \xi) \longmapsto \sum_{i=1}^{n} f(\xi_i)(t_i - t_{i-1}) \in \mathbb{R}. \tag{4.4.5}$$

Show that this net converges iff f is Riemann integrable. What is the limit in this case?

(iii) Assume in addition that f is continuous. Show that in this case f is Riemann integrable. Show also that there are *subsequences* of this net which converge to the limit.

(iv) Use Exercise 4.4.5 to formulate and to prove a version of Fubini's Theorem for Riemann integrals of continuous functions of two variables over a rectangle.

Exercise 4.4.8 (Filter basis) Consider a subset $\mathfrak{B} \subseteq 2^M$ of the power set of some non-empty set M.

(i) Show that $\mathfrak{B}$ is the basis of a (uniquely determined) filter $\mathfrak{F}(\mathfrak{B})$ iff for $A, B \in \mathfrak{B}$ there is a subset $C \subseteq A \cap B$ with $C \in \mathfrak{B}$ and $\emptyset \notin \mathfrak{B}$.

(ii) Show that there is a (uniquely determined) coarsest filter $\mathfrak{F}$ with $\mathfrak{B} \subseteq \mathfrak{F}$ iff for all $n \in \mathbb{N}$ and all subsets $A_1, \dots, A_n \in \mathfrak{B}$ the intersection $A_1 \cap \cdots \cap A_n$ is non-empty. In this case, the filter $\mathfrak{F}$ is called the filter generated by $\mathfrak{B}$ and $\mathfrak{B}$ is called a *subbasis* of $\mathfrak{F}$.

Exercise 4.4.9 (Push-Forward of a filter) Let $f : M \longrightarrow N$ be a map between two sets and let $\mathfrak{F}$ be a filter on M.

(i) Show that $f_*\mathfrak{F}$ is indeed a filter on N.

(ii) Show that the map f_* satisfies the properties of a push-forward, i.e. for $g : L \longrightarrow M$ we have

$$(f \circ g)_* = f_* \circ g_* \quad \text{and} \quad (\mathrm{id}_M)_* = \mathrm{id}. \tag{4.4.6}$$

(iii) How do filter bases behave under push-forward?

Exercise 4.4.10 (Trace filter) Consider a non-empty subset $A \subseteq M$ of a set M and a filter $\mathfrak{F}$ on M.

(i) Show that

$$\mathfrak{F}_A = \{A \cap B \mid B \in \mathfrak{F}\} \tag{4.4.7}$$

defines a filter on A iff $B \cap A \neq \emptyset$ for all $B \in \mathfrak{F}$. In this case the filter $\mathfrak{F}_A$ is called the *trace filter* of $\mathfrak{F}$ on A.

(ii) Assume in addition that $\mathfrak{F}$ is an ultrafilter. Show that $\mathfrak{F}_A$ is a filter on A iff $A \in \mathfrak{F}$. Show that in this case also $\mathfrak{F}_A$ is an ultrafilter.

Exercise 4.4.11 (Accumulation points and cluster points) Consider a topological space $(M, \mathcal{M})$ and a sequence $(p_n)_{n \in \mathbb{N}}$ in M. The direct analogy of the notion of an accumulation point as known from elementary calculus is then, that $p \in M$ is called *accumulation point* of the sequence $(p_n)_{n \in \mathbb{N}}$ if for every neighbourhood $U \in \mathfrak{U}(p)$ of p there are infinitely many points of the sequence in U.

(i) Show that $p \in M$ is an accumulation point of the sequence $(p_n)_{n \in \mathbb{N}}$ iff p is a cluster point of the net $(p_n)_{n \in \mathbb{N}}$.

(ii) Let $(p_i)_{i \in I}$ be a net in M and denote the set of cluster points of it by C. Show that

$$C = \bigcap_{i \in I} A_i, \tag{4.4.8}$$

where $A_i = \{p_j \mid j \succcurlyeq i\}^{\mathrm{cl}}$.

Chapter 5
Compactness

In this section we come to one of the most important concepts in topology: compactness. In fact, there are several competing definitions and approaches which generalize the naive idea of compact subsets of $\mathbb{R}$ as known from calculus. We first note that the definition of compact subsets in $\mathbb{R}$ as being the closed and bounded subsets clearly refers to the metric structure and, hence, will not be available for general topological spaces directly. The property that a subset A of $\mathbb{R}$ is compact if every sequence in A has a convergent subsequence seems more suitable after passing to either general nets and subnets or filters and finer filters. However, the traditional approach is via the covering property of compact subsets: a subset A of $\mathbb{R}$ is compact iff every open cover has a finite subcover. We will now investigate the relations between these different formulations and establish several basic properties of compact spaces. A fundamental theorem on compact spaces is Tikhonov's Theorem which we will prove using ultrafilters. In a next step one relaxes compactness to local compactness which provides many interesting non-compact topological spaces still sharing some nice features with their compact colleagues.

5.1 Compact Spaces

In the definition of a compact topological space there are two conventions in the literature: either the Hausdorff property is explicitly required, like in [3], or not, like in [17]. We follow the second choice:

Definition 5.1.1 (*Compactness*) A topological space $(M, \mathcal{M})$ is called compact if every open cover $\{\mathcal{O}_i\}_{i \in I}$ of M has a finite subcover. A subset $K \subseteq M$ of $(M, \mathcal{M})$ is called compact if $(K, \mathcal{M}|_K)$ is compact.

Recall that an open cover $\{\mathcal{O}_i\}_{i \in I}$ is a collection of open subsets $\mathcal{O}_i \in \mathcal{M}$ of M such that their union is M, i.e.

$$M = \bigcup_{i \in I} \mathcal{O}_i. \tag{5.1.1}$$

© Springer International Publishing Switzerland 2014
S. Waldmann, *Topology*, DOI 10.1007/978-3-319-09680-3_5

Hence compactness means that for any such cover, we find $i_1, \ldots, i_n \in I$ with $M = \mathcal{O}_{i_1} \cup \cdots \cup \mathcal{O}_{i_n}$. For a subset K compactness means that for any collection of open subsets $\{\mathcal{O}_i\}_{i \in I}$ of M with

$$K \subseteq \bigcup_{i \in I} \mathcal{O}_i, \tag{5.1.2}$$

we find finitely many indices $i_1, \ldots, i_n \in I$ with $K \subseteq \mathcal{O}_{i_1} \cup \cdots \cup \mathcal{O}_{i_n}$. This is clear by the definition of the subspace topology.

By taking complements of open subsets we get immediately the following equivalent characterization of compactness.

Proposition 5.1.2 *Let* $(M, \mathcal{M})$ *be a topological space. Then* $(M, \mathcal{M})$ *is compact iff for every collection* $\{A_i\}_{i \in I}$ *of closed subsets* $A_i \subseteq M$ *with* $\bigcap_{i \in I} A_i = \emptyset$ *one finds finitely many* $i_1, \ldots, i_n \in I$ *with* $A_{i_1} \cap \cdots \cap A_{i_n} = \emptyset$.

The next characterization is based on filters, cluster points, and ultrafilters:

Proposition 5.1.3 *Let* $(M, \mathcal{M})$ *be a topological space. The following statements are equivalent:*

(i) *The space* $(M, \mathcal{M})$ *is compact.*
(ii) *Every filter on* M *has a cluster point.*
(iii) *Every ultrafilter on* M *converges.*

Proof First suppose that $(M, \mathcal{M})$ is compact and let $\mathfrak{F}$ be a filter. Recall that a point $p \in M$ is a cluster point if for every $U \in \mathfrak{U}(p)$ and all $A \in \mathfrak{F}$ one has $A \cap U \ne \emptyset$. This means that $p \in A^{\mathrm{cl}}$ for all $A \in \mathfrak{F}$ or $p \in \bigcap_{A \in \mathfrak{F}} A^{\mathrm{cl}}$. Thus the set of cluster points of $\mathfrak{F}$ is given by $\bigcap_{A \in \mathfrak{F}} A^{\mathrm{cl}}$. If this is empty then by Proposition 5.1.2 we find finitely many $A_1, \ldots, A_n \in \mathfrak{F}$ with $A_1^{\mathrm{cl}} \cap \cdots \cap A_n^{\mathrm{cl}} = \emptyset$ and hence also $A_1 \cap \cdots \cap A_n = \emptyset$. But this can not be the case for a filter, yielding a contradiction. Next assume (ii) and let $\mathfrak{F}$ be an ultrafilter. Thus $\mathfrak{F}$ has a cluster point, to which $\mathfrak{F}$ converges by Proposition 4.3.3. Finally, assume (iii) and let $\{\mathcal{O}_i\}_{i \in I}$ be an open cover of $(M, \mathcal{M})$ for which there is *no* finite subcover. Then the sets $A_J = M \setminus \bigcup_{i \in J} \mathcal{O}_i$ for finite subsets $J \subseteq I$ provide a family of subsets of M with $A_J \cap A_{J'} \ne \emptyset$ for all finite subsets J, J'. Thus they form a basis of a filter $\mathfrak{F}$ which, by Proposition 4.3.1, is contained in an ultrafilter $\mathfrak{G}$. By assumption, $\mathfrak{G}$ converges to some point $p \in M$. Since the $\{\mathcal{O}_i\}_{i \in I}$ form a cover, there is a $i \in I$ with $p \in \mathcal{O}_i$. Hence $\mathcal{O}_i$ is a neighbourhood of p and by convergence of $\mathfrak{G}$ this implies $\mathcal{O}_i \in \mathfrak{G}$. However, $A_J \in \mathfrak{G}$ implies $M \setminus A_J \notin \mathfrak{G}$ for all J and thus $\mathcal{O}_i = M \setminus A_{\{i\}} \notin \mathfrak{G}$, a contradiction. $\square$

Since $p \in M$ is a cluster point of a filter iff p is a cluster point of the associated net and since p is a cluster point of a net iff it is a cluster point of the associated filter, we get the following re-formulation of the last proposition:

Proposition 5.1.4 *Let* $(M, \mathcal{M})$ *be a topological space. Then* $(M, \mathcal{M})$ *is compact iff every net in* M *has a convergent subnet.*

Proof This is now a consequence of Propositions 4.2.6 and 5.1.3.   □

Remark 5.1.5 This statement reminds very much on the compactness used in calculus based on the Bolzano-Weierstraß Theorem, i.e. a subset K of $\mathbb{R}^n$ is compact iff every sequence in K has a convergent subsequence in K. However, in the present general situation, we can conclude only the following: for a sequence in a compact space there exists a convergent subnet of this sequence. But this subnet needs *not* to be a subsequence at all. Conversely, if every sequence in a topological space has a convergent subsequence, this does not yet imply compactness as in Proposition 5.1.4 we need general nets and not just sequences.

These subtleties motivate now the following definition:

Definition 5.1.6 (*Sequential Compactness*) A topological space $(M, \mathcal{M})$ is called sequentially compact if every sequence in M has a convergent subsequence.

Various examples show that in general, compactness and sequential compactness are quite unrelated. Nevertheless, there are favorable situations where both notions coincide. We will come back to this in Sect. 5.4.

The next characterization of compact spaces is extremely useful as it only refers to a subbasis of the topology instead of the topology itself:

Theorem 5.1.7 (Alexander) *Let $(M, \mathcal{M})$ be a topological space and let $\mathcal{S} \subseteq \mathcal{M}$ be a subbasis of the topology. Then $(M, \mathcal{M})$ is compact iff every cover of M by subsets from $\mathcal{S}$ has a finite subcover.*

Proof Only one direction is non-trivial: assume that every cover by subsets of $\mathcal{S}$ has a finite subcover and assume that M is non-compact. Thus there is an ultrafilter $\mathfrak{F}$ on M which does not converge by the characterization of compactness via ultrafilters as in Proposition 5.1.3. Hence for every point $p \in M$ we have a neighbourhood $U_p \in \mathfrak{U}(p)$ with $U_p \notin \mathfrak{F}$. Since $\mathcal{S}$ is a subbasis, we can assume U_p is a finite intersection of elements in $\mathcal{S}$. Since $p \in U_p$, every of these open subsets in $\mathcal{S}$ has to contain p. Hence we can assume $p \in U_p \in \mathcal{S}$ from the beginning. This gives an open cover $\{U_p\}_{p \in M}$ of M by subsets from $\mathcal{S}$. By assumption, we find $p_1, \ldots, p_n \in M$ with $U_{p_1} \cup \cdots \cup U_{p_n} = M$. Since $U_p \notin \mathfrak{F}$ we have $M \setminus U_p \in \mathfrak{F}$ by Proposition 4.3.2, (i). Since $\mathfrak{F}$ is a filter, also $(M \setminus U_{p_1}) \cap \cdots \cap (M \setminus U_{p_n}) \in \mathfrak{F}$ but this intersection is empty as $U_{p_1} \cup \cdots \cup U_{p_n} = M$, contradicting the filter properties. Hence M is compact.   □

As a first simple application of Alexander's Theorem we obtain the compactness of the closed intervals $[a, b]$ in $\mathbb{R}$:

Example 5.1.8 Consider a closed interval $[a, b] \subseteq \mathbb{R}$ with its usual subspace topology. Then it is easy to see that the open subsets $[a, c)$ and $(c', b]$ with $c, c' \in [a, b]$ constitute a subbasis. Assume that we have an open cover U by subsets from this subbasis. Define $x = \sup\{c \mid [a, c) \in U\}$. Note that there are necessarily subsets $[a, c) \in U$ as well as $(c', b] \in U$ as otherwise one has no cover. To have a cover, we have a $c' < x$ with $(c', b] \in U$. By the definition of the supremum we also have a $c > c'$ with $[a, c) \in U$. Thus $[a, c)$ and $(c', b]$ is a finite subcover of U. Hence $[a, b]$ is compact by Alexander's Theorem.

We conclude this section with some further properties of compact spaces.

Proposition 5.1.9 *Let* $(M, \mathcal{M})$ *be compact and let* $A \subseteq M$ *be closed. Then* $(A, \mathcal{M}|_A)$ *is compact, too.*

Proof Since the closed subsets of A in the subspace topology are precisely the closed subsets $B \subseteq M$ with $B \subseteq A$, see Exercise 2.7.3, we get the result immediately by Proposition 5.1.2.   □

The converse needs not to be true as the following example shows:

Example 5.1.10 Let M be a set with at least 2 elements and consider $\mathcal{M}_{\text{indiscrete}}$, i.e. the indiscrete topology. Since $\emptyset$ and M are the only open subsets, any open cover has to contain M itself. Thus $(M, \mathcal{M}_{\text{indiscrete}})$ is compact. Since there are non-trivial subsets there are compact but non-closed subsets of this topological space.

If the topological space is in addition Hausdorff, things will behave nicer: we have the desired equivalence. Moreover, we get an additional separation property for free:

Proposition 5.1.11 *Let* $(M, \mathcal{M})$ *be a Hausdorff space.*

 (i) *If* $K \subseteq M$ *is compact and* $p \in M \setminus K$ *then there exist open subsets* $\mathcal{O}_1, \mathcal{O}_2 \subseteq M$ *with* $\mathcal{O}_1 \cap \mathcal{O}_2 = \emptyset$ *and* $p \in \mathcal{O}_1$, $K \subseteq \mathcal{O}_2$.
 (ii) *Every compact subset* $K \subseteq M$ *is closed.*
 (iii) *If* $(M, \mathcal{M})$ *is compact then* $(M, \mathcal{M})$ *is regular.*

Proof By the Hausdorff property we find for every $q \in K$ disjoint open subsets $\mathcal{O}_q \in \mathfrak{U}(q)$ and $U_q \in \mathfrak{U}(p)$ to separate q from p. Then $K \subseteq \bigcup_{q \in K} \mathcal{O}_q$ and hence $\{\mathcal{O}_q\}_{q \in K}$ is an open cover of K. By compactness there are $q_1, \ldots, q_n \in K$ with $K \subseteq \mathcal{O}_{q_1} \cup \cdots \cup \mathcal{O}_{q_n} = \mathcal{O}_2$. Moreover $\mathcal{O}_1 = U_{q_1} \cap \cdots \cap U_{q_n}$ is still an open neighbourhood of p and $\mathcal{O}_1 \cap \mathcal{O}_2 = \emptyset$, showing the first part. For the second part, choose open subsets $\mathcal{O}_p \subseteq M$ for every $p \in M \setminus K$ with $\mathcal{O}_p \cap K = \emptyset$, according to the first part. Then $M \setminus K = \bigcup_{p \in M \setminus K} \mathcal{O}_p$ is open. The last part is now clear as every closed subset of M is compact and can be separated from a disjoint point according to the first part.   □

Corollary 5.1.12 *A subset* $K \subseteq \mathbb{R}$ *is compact iff it is closed and bounded.*

Proof If K is closed and bounded it is a closed subset of a compact interval, hence compact itself by Proposition 5.1.9. Conversely, the sets $(-n, n)$ for $n \in \mathbb{N}$ give an open cover of $\mathbb{R}$ and hence a compact subset K has to be contained in one of them. Thus K is bounded. By Proposition 5.1.11, (ii), it is also closed.   □

The last part of Proposition 5.1.11 can be sharpened to the following result:

Proposition 5.1.13 *Let $(M, \mathcal{M})$ be a compact Hausdorff space. Then $(M, \mathcal{M})$ is normal.*

Proof In order to check T_4 we take two closed disjoint subsets $A, B \subseteq M$ which are compact themselves by Proposition 5.1.9. For every $p \in A$ we find an open neighbourhood $U_p \in \mathfrak{U}(p)$ and an open subset $V_p \subseteq M$ with $U_p \cap V_p = \emptyset$ and $B \subseteq V_p$. By compactness, finitely many points $p_1, \ldots, p_n$ will suffice to give a cover $A \subseteq U_{p_1} \cup \cdots \cup U_{p_n} = \mathcal{O}_1$. Then the finite intersection $\mathcal{O}_2 = V_{p_1} \cap \cdots \cap V_{p_n}$ is still open and covers B. Clearly $\mathcal{O}_1 \cap \mathcal{O}_2 = \emptyset$. $\qquad\square$

5.2 Continuous Maps and Compactness

In this short section we discuss how compact spaces behave under continuous maps.

Proposition 5.2.1 *Let $f : (M, \mathcal{M}) \longrightarrow (N, \mathcal{N})$ be a continuous map between topological spaces and let $K \subseteq M$ be compact. Then $f(K) \subseteq N$ is compact, too.*

Proof Let $\{\mathcal{O}_i\}_{i \in I}$ be an open cover of $f(K)$ by some open subsets $\mathcal{O}_i \in \mathcal{N}$. Then $f^{-1}(\mathcal{O}_i)$ is open by the continuity of f and $K \subseteq \bigcup_{i \in I} f^{-1}(\mathcal{O}_i)$. Hence finitely many $f^{-1}(\mathcal{O}_{i_1}), \ldots, f^{-1}(\mathcal{O}_{i_n})$ cover the compact subset K. But

$$f(K) \subseteq f(f^{-1}(\mathcal{O}_{i_1}) \cup \cdots \cup f^{-1}(\mathcal{O}_{i_n})) \subseteq \mathcal{O}_{i_1} \cup \cdots \cup \mathcal{O}_{i_n}$$

gives then the finite open subcover of $f(K)$. $\qquad\square$

Note that compact subsets behave just the opposite to open or closed subsets: images of compact subsets are again compact, but preimages need not to be compact at all:

Example 5.2.2 Let $(M, \mathcal{M})$ be a non-compact topological space and let $f : M \longrightarrow \{pt\}$ be the unique map sending M to the one-point space. For the unique topology on $\{pt\}$ this is clearly continuous and $\{pt\}$ is compact. The image of every subset of M is compact but $f^{-1}(\{pt\}) = M$ is *not* compact.

Definition 5.2.3 (*Proper map*) Let $f : (M, \mathcal{M}) \longrightarrow (N, \mathcal{N})$ be a continuous map between topological spaces. Then f is called proper if $f^{-1}(K) \subseteq M$ is compact for every compact $K \subseteq N$.

In particular, the map from Example 5.2.2 is not proper. Properness becomes interesting whenever the domain is non-compact. In fact, we have the following statement:

Proposition 5.2.4 *Let $f : (M, \mathcal{M}) \longrightarrow (N, \mathcal{N})$ be a continuous map between topological spaces where $(M, \mathcal{M})$ is compact and $(N, \mathcal{N})$ is Hausdorff. Then f is proper.*

Proof Let $K \subseteq N$ be compact then K is closed by Proposition 5.1.11, (ii). Thus $f^{-1}(K) \subseteq M$ is closed by continuity of f and hence compact by Proposition 5.1.9. □

The next statement turns out to be very useful for showing the homeomorphism property for maps between compact spaces:

Proposition 5.2.5 *Let* $f: (M, \mathcal{M}) \longrightarrow (N, \mathcal{N})$ *be a continuous map from a compact space to a Hausdorff space.*

(i) The map f is closed.
(ii) If f is injective then f is an embedding.
(iii) If f is bijective then f is a homeomorphism.

Proof Let $A \subseteq M$ be closed then A is compact by Proposition 5.1.9. Hence $f(A) \subseteq N$ is compact by Proposition 5.2.1 and thus closed by Proposition 5.1.11, (ii). This shows the first part. Now let f be injective in addition. Then f is a closed injective map according to the first part. Then also the map $f: (M, \mathcal{M}) \longrightarrow (f(M), \mathcal{N}|_{f(M)})$ is a closed map since $f(A) \subseteq f(M)$ is clearly closed in the subspace topology once $f(A) \subseteq N$ is closed. Hence f is a homeomorphism onto its image by Proposition 2.4.8, (iii). Then the third is clear. □

The last result in this section is familiar from calculus again:

Proposition 5.2.6 *Let* $(M, \mathcal{M})$ *be a compact space and* $f \in \mathscr{C}(M, \mathbb{R})$. *Then f is bounded and there are points $p_{\max}$ and $p_{\min}$ in M with*

$$\sup\{f(p) \mid p \in M\} = f(p_{\max}) \tag{5.2.1}$$

and

$$\inf\{f(p) \mid p \in M\} = f(p_{\min}). \tag{5.2.2}$$

Proof We know that $f(M)$ is compact in $\mathbb{R}$ hence bounded. Thus for all $q \in M$ we have

$$-\infty < \inf\{f(p) \mid p \in M\} \leq f(q) \leq \sup\{f(p) \mid p \in M\} < +\infty.$$

Since a compact subset of $\mathbb{R}$ is also closed, we find points $p_{\min}$ and $p_{\max}$ realizing the inf and sup as a min and as a max, respectively. □

Note that if in addition M is connected then we get from Proposition 2.5.4 the result

$$f(M) = \big[f(p_{\min}), f(p_{\max})\big]. \tag{5.2.3}$$

Without connectedness, we can have a union of compact intervals.

5.3 Tikhonov's Theorem

We come now to one of the most important theorems in general topology: Tikhonov's Theorem states that the Cartesian product of compact spaces is again compact. The applications of this theorem go far beyond topology itself. In functional analysis many fundamental results rely on this like the Banach-Alaoglu theorem, the Gel'fand-Naimark theory of commutative C^*-algebras, and many more. Interesting enough, Tikhonov's Theorem in its general form is equivalent to the Axiom of Choice, see e.g. [11] for a detailed discussion of the role of the Axiom of Choice in topology. We come now to the precise formulation of the theorem and its proof. With our present machinery this is almost disappointingly easy: note however, that we just harvest many non-trivial investments done before.

Theorem 5.3.1 (Tikhonov) *Let I be an arbitrary non-empty index set and let $(M_i, \mathcal{M}_i)_{i \in I}$ be a family of non-empty topological spaces. Then their Cartesian product $M = \prod_{i \in I} M_i$ with the product topology is compact iff each $(M_i, \mathcal{M}_i)$ is compact.*

Proof Suppose first that M is compact. Then $M_i = \mathrm{pr}_i(M)$ is the image of a compact topological space under a continuous map according to Proposition 3.1.2, (i). Hence Proposition 5.2.1 shows that $(M_i, \mathcal{M}_i)$ is compact, too. Conversely, let $(M_i, \mathcal{M}_i)_{i \in I}$ be compact for each $i \in I$. If $\mathfrak{F}$ is an ultrafilter on M then the image filters $\mathfrak{F}_i = (\mathrm{pr}_i)_* \mathfrak{F}$ are ultrafilters in M_i for each $i \in I$ by Proposition 4.3.2, (iii). From Proposition 5.1.3 we conclude that $\mathfrak{F}_i$ converges for each $i \in I$, say to $p_i \in M_i$. Then Exercise 4.4.4 shows that also $\mathfrak{F}$ converges, namely to $p = (p_i)_{i \in I}$. Hence Proposition 5.1.3 shows that $(M, \mathcal{M})$ is compact, too. $\square$

A slightly different proof using Alexander's Theorem is discussed in Exercise 5.5.4. The validity of Tikhonov's Theorem relies very much on the correct definition of the product topology. In particular, the seemingly easier *box topology* from Exercise 3.4.4 will not be suitable, see Exercise 5.5.5.

A first important application is obtained from Tikhonov's Theorem together with the results of Proposition 3.1.2, (iv).

Corollary 5.3.2 *Let $(M_i, \mathcal{M}_i)_{i \in I}$ be as in Theorem 5.3.1. Then M is a compact Hausdorff space iff $(M_i, \mathcal{M}_i)$ is a compact Hausdorff space for all $i \in I$.*

Another immediate and simple application of Tikhonov's Theorem, not using its full strength yet, is given by the Heine-Borel Theorem known from elementary calculus.

Theorem 5.3.3 (Heine-Borel) *A subset $K \subseteq \mathbb{R}^n$ is compact iff it is bounded and closed.*

Proof Using the open balls $B_k(0) \subseteq \mathbb{R}^n$ instead of open intervals $(-k, k) \in \mathbb{R}$ for $k \in \mathbb{N}$ shows analogously to Corollary 5.1.12 that a compact subset $K \subseteq \mathbb{R}^n$ is bounded and closed. The converse follows now from Example 5.1.8 and Tikhonov's Theorem since every bounded subset is in some large but compact cube $[-k, k]^n \subseteq \mathbb{R}^n$. $\square$

5.4 Further Notions of Compactness

We discuss now several further notions generalizing compactness: the first observation is that e.g. $\mathbb{R}^n$ is not compact but shares still many nice features of compact
spaces: locally around a point it still looks like being compact since we have compact
neighbourhoods of every point p, the closed balls $B_r(p)^{cl} \subseteq \mathbb{R}^n$. This is enough for
many conclusions and motivates the following definition:

Definition 5.4.1 (*Local compactness*) A topological space $(M, \mathcal{M})$ is called locally
compact if every point $p \in M$ has a compact neighbourhood.

Obviously, a compact space is locally compact since we can take the whole space
as a compact neighbourhood. Note also that some authors include the Hausdorff
property in the definition of local compactness like [3, 27] but not [17]. In combination
with the Hausdorff property we get a stronger separation property for locally compact
spaces:

Proposition 5.4.2 *Let* $(M, \mathcal{M})$ *be a locally compact Hausdorff space. Then* $(M, \mathcal{M})$
is regular.

Proof Let $p \in M$ and let $K \in \mathfrak{U}(p)$ be a compact neighbourhood of p. Since M is
Hausdorff, K is closed by Proposition 5.1.11, (ii). Now let $\mathcal{O} \subseteq M$ be an arbitrary
open neighbourhood of p then $\mathcal{O} \cap K$ is a neighbourhood of p viewed as point in
M *and* viewed as point in K. By Proposition 2.6.5 applied to the T_3-space K we
find an open neighbourhood $U \subseteq K$ of p with $U \subseteq U^{cl} \subseteq \mathcal{O} \cap K$. Since K is a
neighbourhood of p in M, too, the subset $U \subseteq M$ is also a neighbourhood of p in
M. Since $K = K^{cl}$ is closed in M, the closure $\overline{U}$ with respect to $\mathcal{M}\big|_K$ coincides
with the closure U^{cl} with respect to $\mathcal{M}$. Hence we have found a neighbourhood U
of p with $U \subseteq U^{cl} \subseteq \mathcal{O}$. Shrinking U again shows that we can assume U to be open
in $(M, \mathcal{M})$. Hence Proposition 2.6.5 applies again and $(M, \mathcal{M})$ is a T_3-space. □

Corollary 5.4.3 *Let* $(M, \mathcal{M})$ *be a locally compact Hausdorff space and let* $p \in M$.
Then there exists a neighbourhood basis of compact neighbourhoods.

Proof Since $(M, \mathcal{M})$ is regular by Proposition 5.4.2 we get a neighbourhood basis
$\{A_i\}_{i \in I}$ of closed neighbourhoods of p from Proposition 2.6.5. If $K \subseteq M$ is a
compact neighbourhood, then $\{A_i \cap K\}_{i \in I}$ is a neighbourhood basis consisting of
compact subsets. □

Many of the Hausdorff space we have seen turn out to be locally compact, even
though they are non necessarily compact:

Example 5.4.4 (*Locally compact spaces*)

 (i) As already mentioned every compact space is also locally compact.
(ii) Any open subset of a locally compact Hausdorff space is again locally compact
     Hausdorff.

(iii) $\mathbb{R}^n$ with its usual topology is locally compact since $B_R(p)^{\mathrm{cl}}$ is a compact neighbourhood of p by the Heine-Borel Theorem.

(iv) Any topological manifold is locally compact. Indeed, we can decide local compactness inside a small open neighbourhood of a given point. There, a topological manifold is homeomorphic to an open subset of $\mathbb{R}^n$ and hence locally compact.

Further properties and constructions of locally compact spaces will be discussed in Exercise 5.5.8 as well as in Exercise 6.4.3.

The next notion of compactness deals with more particular open covers:

Definition 5.4.5 (*Countable compactness*) A topological space $(M, \mathcal{M})$ is called countably compact if every countable open cover of M has a finite subcover.

Proposition 5.4.6 *Let $(M, \mathcal{M})$ be a topological space. Then the following statements are equivalent:*

 (i) The space $(M, \mathcal{M})$ is countably compact.
 (ii) Every countable collection of closed subsets of M with empty intersection has a finite subset with empty intersection.
(iii) Every sequence in M has an accumulation point.

Proof The equivalence of (i) and (ii) is clear by taking complements. Thus assume (ii) and let $(p_n)_{n \in \mathbb{N}}$ be a sequence. Recall that $p \in M$ is an *accumulation point* of $(p_n)_{n \in \mathbb{N}}$ if every neighbourhood $U \in \mathfrak{U}(p)$ of p contains infinitely many p_n, see also Exercise 4.4.11, (i). Suppose that $(p_n)_{n \in \mathbb{N}}$ has no accumulation points. Then consider $P_n = \{p_m \mid m \geq n\} \subseteq M$ and $A_n = P_n^{\mathrm{cl}}$, the closure of the end piece of the sequence. We know that the set of accumulation points of $(p_n)_{n \in \mathbb{N}}$ is given by $\bigcap_{n=1}^{\infty} A_n$, which is empty by assumption, see Exercise 4.4.11, (ii). From (ii) we get finitely many $A_{n_1}, \ldots, A_{n_k}$ with $A_{n_1} \cap \cdots \cap A_{n_k} = \emptyset$. However, $A_{n_1} \cap \cdots \cap A_{n_k}$ always contains $P_{n_1} \cap \cdots \cap P_{n_k} = \{p_m \mid m \geq \max\{n_1, \ldots, n_k\}\}$ and thus we reached a contradiction. Conversely, assume (iii) and let $\{\mathcal{O}_n\}_{n \in \mathbb{N}}$ be a countable open cover with no finite subcover. Thus $M \setminus (\mathcal{O}_1 \cup \cdots \cup \mathcal{O}_n)$ is never empty and we find $p_n \in M \setminus (\mathcal{O}_1 \cup \cdots \cup \mathcal{O}_n)$. This defines a sequence $(p_n)_{n \in \mathbb{N}}$ which has an accumulation point $p \in M$ by assumption. Since $\{\mathcal{O}_n\}_{n \in \mathbb{N}}$ is a cover, we have $p \in \mathcal{O}_N$ for some $N \in \mathbb{N}$. But $p_n \in M \setminus (\mathcal{O}_1 \cup \cdots \cup \mathcal{O}_n)$ for all $n > N$ and thus the open neighbourhood $\mathcal{O}_N$ of p can contain at most the finitely many $p_1, \ldots, p_N$. Thus we reached a contradiction also for this assertion. $\square$

Again, a warning is appropriate: for an accumulation point p of a sequence (which is the same as a cluster point) we do not get a subsequence converging to p in general, only a *subnet*, see Exercise 4.4.6. If however, we have such a convergent subsequence for every sequence, then a limit of a subsequence is an accumulation point of the sequence. This shows the following important implication.

Proposition 5.4.7 *A sequentially compact space is countably compact.*

The reason why the converse implication may fail is governed by the first countability axiom. In fact, for first countable spaces we get the equivalence of sequential and countable compactness.

Proposition 5.4.8 *Let $(M, \mathcal{M})$ be a topological space satisfying the first countability axiom. Then $(M, \mathcal{M})$ is countably compact iff it is sequentially compact.*

Proof Let M be countably compact and let $(p_n)_{n \in \mathbb{N}}$ be a sequence with accumulation point $p \in M$ according to Proposition 5.4.6, (iii). Choose a countable neighbourhood basis $\{U_m\}_{m \in \mathbb{N}}$ of this point p. Without restriction we can assume $U_{m+1} \subseteq U_m$ for all $m \in \mathbb{N}$. The point p being an accumulation point gives us an index $n_{m+1} > n_m$ for every $m \in \mathbb{N}$ with $p_{n_m} \in U_m$. Then this subsequence $(p_{n_m})_{m \in \mathbb{N}}$ converges to p, showing that $(M, \mathcal{M})$ is sequentially compact. The converse holds in general thanks to Proposition 5.4.7.   □

We can also relate countable compactness to compactness in the case of a space which is second countable. To this end we first state Lindelöf's Theorem.

Theorem 5.4.9 (Lindelöf) *Let $(M, \mathcal{M})$ be a topological space which is second countable and let $N \subseteq M$ be a subset. Then every open cover of N has a countable subcover.*

Proof Clearly, we can assume $M = N$. Let $\{\mathcal{O}_i\}_{i \in I}$ be an open cover and let $\{U_n\}_{n \in \mathbb{N}}$ be a countable basis of the topology. Then for every $i \in I$ we have

$$\mathcal{O}_i = \bigcup_{j \in J_i} U_j$$

with some appropriate $J_i \subseteq \mathbb{N}$. But then $\bigcup_{i \in I} \bigcup_{j \in J_i} U_j$ covers M. Define

$$J = \bigcup_{i \in I} J_i \subseteq \mathbb{N},$$

which is at most countable. For every $n \in J$ there exists at least one $i \in I$ with $U_n \subseteq \mathcal{O}_i$. Choose such a $i_n \in I$ for every $n \in J$ then one gets a countable subcover $\{\mathcal{O}_{i_n}\}_{n \in J}$.   □

Corollary 5.4.10 *For a second countable space the notions of compactness, countable compactness, and sequential compactness coincide.*

Proof A second countable space is in particular first countable. The equivalence of compactness and countable compactness follows from Lindelöf's Theorem, the equivalence of countable compactness and sequential compactness is the content of Proposition 5.4.8.   □

Corollary 5.4.11 *For subsets of $\mathbb{R}^n$ or, more generally, for subsets of topological manifolds the notions of compactness, countable compactness, and sequential compactness agree.*

More generally, for a metric space all notions of compactness coincide.

Proposition 5.4.12 *Let (M, d) be a metric space. Then the following statements are equivalent:*

(i) The space (M, d) is compact.
(ii) The space (M, d) is countably compact.
(iii) The space (M, d) is sequentially compact.

Proof Since (M, d) is first countable we have (ii) $\Longleftrightarrow$ (iii) according to Proposition 5.4.8. Moreover, (i) $\Longrightarrow$ (ii) is true in general. Thus assume that M is countably compact. Let $k \geq 1$ then there is *no* infinite subset $N_k \subseteq M$ with

$$d(p, q) \geq \frac{1}{k}$$

for all $p, q \in N_k$ with $p \neq q$. Indeed, if there would be an infinite subset N_k with this property we could choose a sequence $(p_n)_{n \in \mathbb{N}}$ inside N_k of pairwise distinct points having all distance $\geq \frac{1}{k}$. But such a sequence has clearly no accumulation points, contradicting (iii). Thus there exists a *finite* subset $A_k \subseteq M$ with the property that for all $p \in M$ there is a $q \in A_k$ with $d(p, q) < \frac{1}{k}$. Indeed, suppose this is not the case then for all finite subsets A of M there is a point p with $d(p, q) \geq \frac{1}{k}$ for all $q \in A$. Now start with some point $\{q_1\}$ and find q_2 with $d(q_2, q_1) \geq \frac{1}{k}$. Since again the set $\{q_1, q_2\}$ is finite, by induction we find points $q_1, q_2, \ldots$, with pairwise distance at least $\frac{1}{k}$, hence constituting a subset N_k as above. But this we showed not to exist. Thus we have a contradiction and we can conclude the existence of a finite subset A_k as wanted. Then this implies that the countable set $A = \bigcup_{k=1}^{\infty} A_k \subseteq M$ is *dense* in M. Hence we have found a countable dense subset in a metric space. Taking the open balls around the points in A with rational radii gives a countable basis of the topology by same argument as for $\mathbb{R}^n$ in Example 2.3.7. Thus Lindelöf's Theorem applies and from Corollary 5.4.10 we get the remaining implication (ii) $\Longrightarrow$ (i). $\square$

The following result was obtained in course of the above proof and is of independent interest.

Corollary 5.4.13 *Let (M, d) be a compact metric space. Then there exists a countable dense subset of M.*

There are yet more notions of compactness: two of them will be of great importance in (differential) geometry. We will discuss them in Sect. 6.2 since at the moment their definition would seem quite unmotivated.

5.5 Exercises

Exercises 5.5.1 (The Cantor set II) Consider again the Cantor set C from Exercise 2.7.24.

(i) Show that C is compact.
(ii) Consider the Cartesian product $M = \{0, 2\}^{\mathbb{N}}$ of countably many copies of the discrete topological space with two points $\{0, 2\}$. Let $f : M \longrightarrow C$ be defined by

$$f(a) = \sum_{n=1}^{\infty} \frac{a_n}{3^n}. \tag{5.5.1}$$

Show that f is continuous and bijective.
(iii) Show that also f^{-1} is continuous. This way, C is homeomorphic to the Cartesian product $\{0, 2\}^{\mathbb{N}}$ endowed with the product topology.

Exercises 5.5.2 (Compact spaces) Consider a set M endowed with the discrete, the indiscrete, or the cofinal topology. Under which assumptions on M does this yield a compact space?

Exercises 5.5.3 (The diagonal is proper) Let $(M, \mathcal{M})$ be a topological Hausdorff space. Show that the diagonal map $\Delta : M \longrightarrow M \times M$ from Exercise 3.4.2 is a proper map.
Hint: Exercise 3.4.2, (iv), will be helpful.

Exercises 5.5.4 (The Theorem of Tikhonov) Consider a collection of non-empty compact topological spaces $(M_i, \mathcal{M}_i)_{i \in I}$ for an arbitrary non-empty index set I. Endow their Cartesian product $M = \prod_{i \in I} M_i$ with the product topology. Show that in this case M is again compact by using Alexander's Theorem directly.
Hint: The good choice for a subbasis $\mathcal{S}$ of the product topology are of course the open subsets of the form $\mathrm{pr}_i^{-1}(\mathcal{O}_i)$, where $i \in I$ and $\mathcal{O}_i \in \mathcal{M}_i$. To get a proof by contradiction, suppose there is a cover $\mathfrak{U}$ of sets in $\mathcal{S}$, which does not have a finite subcover. Decompose $\mathfrak{U} = \bigcup_{i \in I} \mathfrak{U}_i$ into subsets, where $\mathfrak{U}_i$ contains only subsets of the form $\mathrm{pr}_i^{-1}(\mathcal{O}_i)$ with $\mathcal{O}_i \in \mathcal{M}_i$ for this fixed i. What can one say about the subsets $\mathrm{pr}_i(U_i)$ of M_i for $U_i \in \mathfrak{U}_i$? One needs to choose now in a clever way.

Exercises 5.5.5 (Box topology and Tikhonov's Theorem) Find an example of a countable set of compact topological spaces $(M_n, \mathcal{M}_n)_{n \in \mathbb{N}}$ such that their Cartesian product $M = \prod_{n=1}^{\infty} M_n$ is *not* compact with respect to the box topology from Exercise 3.4.4.
Hint: Exercise 5.5.1 and Exercise 5.5.2.

Exercises 5.5.6 (Union of compact subsets) Let $(M, \mathcal{M})$ be a topological space and let $\{K_i\}_{i \in I}$ be compact subsets of M. Consider their union $K = \bigcup_{i \in I} K_i$ and discuss under which conditions this will be again a compact subset. Find appropriate counterexamples if necessary.

Exercises 5.5.7 (Totally bounded and complete metric spaces) A metric space (M, d) is called *totally bounded*, if for every $\epsilon > 0$ there are finitely many points $p_1, \ldots, p_N \in M$ such that

$$M = \bigcup_{n=1}^{N} B_\epsilon(p_n). \tag{5.5.2}$$

Prove that a metric space is compact iff it is totally bounded and complete.

Hint: The proof of Proposition 5.4.12 should give inspiration for one direction. For the other, assume that there is an open cover $\{\mathcal{O}_i\}_{i \in I}$ without finite subcover. Set $K_0 = M$ and $\epsilon_0 = \frac{1}{2}$. Then there are finitely many points $p_1, \ldots, p_n$ such that the closed balls $B_{1/2}(p_1) \cup \cdots \cup B_{1/2}(p_n)$ cover K_0. These balls have all a diameter of 1. At least one of these balls can not have a finite subcover by the open cover $\{\mathcal{O}_i\}_{i \in I}$. Let K_1 be such a closed ball. Repeating the construction now with K_1 in place of K_0 and $\epsilon_1 = \frac{1}{2}$ etc. one obtains inductively a sequence of closed subsets $K_0 \supseteq K_1 \supseteq \cdots \supseteq K_n \supseteq \cdots$ such that the diameter of K_n is at most $\frac{1}{n}$ but none of the K_n has a finite subcover with respect to the open cover $\{\mathcal{O}_i\}_{i \in I}$. Choose a sequence $(p_n)_{n \in \mathbb{N}}$ with $p_n \in K_n$ and show that p_n converges to some point $p \in \bigcap_{n=0}^{\infty} K_n$. Let $\mathcal{O}_{i_0}$ be one member of the cover with $p \in \mathcal{O}_{i_0}$. Conclude that $K_n \subseteq \mathcal{O}_{i_0}$ for large enough n. This will finally yield a contradiction.

Exercises 5.5.8 (Alexandroff compactification) Let $(M, \mathcal{M})$ be a topological space. Consider the set $M^* = M \cup \{\infty\}$, where one adds just one new point ∞. On M^* one defines a topology $\mathcal{M}^*$, by taking all the open subsets of M and in addition the subsets of the form $M^* \setminus K$ with a closed compact subset $K \subseteq M$.

(i) Show that $(M^*, \mathcal{M}^*)$ is indeed a topological space.
(ii) Determine all open neighbourhoods of ∞ with respect to $\mathcal{M}^*$.
(iii) Show that the subspace topology of $M \subseteq M^*$ coincides with the original topology $\mathcal{M}$ of M.
(iv) Show that $(M^*, \mathcal{M}^*)$ is compact.
(v) Show that $(M^*, \mathcal{M}^*)$ is Hausdorff iff $(M, \mathcal{M})$ is a locally compact Hausdorff space.

The compact space $(M^*, \mathcal{M}^*)$ is called the *Alexandroff compactification* of $(M, \mathcal{M})$.

Exercises 5.5.9 (Compactness of $\mathbb{S}^n$, $\mathbb{T}^n$, $\mathbb{RP}^n$, and $\mathbb{CP}^n$) Let $n \in \mathbb{N}$.

(i) Use Theorem 5.3.3 to show that the n-sphere $\mathbb{S}^n$ is compact.
(ii) Use Theorem 5.3.1 to show that the n-torus $\mathbb{T}^n$ is compact.
(iii) Show that the antipodal map $\mathbb{S}^n \longrightarrow \mathbb{S}^n$ gives a group action of $\mathbb{Z}_2$ by homeomorphisms of $\mathbb{S}^n$. Use this and the ideas from Exercise 3.4.7 to show that the quotient $\mathbb{S}^n / \mathbb{Z}_2$ is homeomorphic to $\mathbb{RP}^n$. Conclude that $\mathbb{RP}^n$ is compact.
(iv) Consider the $(2n + 1)$-sphere $\mathbb{S}^{2n+1}$ as subspace of $\mathbb{C}^{n+1} \setminus \{0\}$. Use the group homomorphism $\mathbb{C} \setminus \{0\} \cong \mathbb{R}^+ \times \mathrm{U}(1)$ and Exercise 3.4.7 to show that there is a group action of $\mathrm{U}(1)$ on $\mathbb{S}^{2n+1}$ by homeomorphisms with the property that the quotient $\mathbb{S}^{2n+1} / \mathrm{U}(1)$ becomes homeomorphic to $\mathbb{CP}^n$. This is the famous *Hopf fibration*. Conclude that also the complex projective space $\mathbb{CP}^n$ is compact.

Exercises 5.5.10 (Proper closed maps) Let $f: (M, \mathcal{M}) \longrightarrow (N, \mathcal{N})$ be a proper map with a locally compact Hausdorff space N. Prove that f is a closed map.

Hint: Use a convergent net $(y_i)_{i \in I}$ to approach a limit point of $f(A)$ for a closed subsets $A \subseteq M$ with $y_i \in f(A)$ and choose preimages $x_i \in A$ of the y_i. For a compact neighbourhood $K \subseteq N$ of $y = \lim_{i \in I} y_i \in f(A)^{\mathrm{cl}}$ consider the compact subset $f^{-1}(K)$.

This is not the most general scenario where proper maps turn out to be closed, see e.g. [26] for a more general statement. Nevertheless, the above case covers a lot of interesting situations.

Exercises 5.5.11 (Locally compact quotients) Consider a locally compact space $(M, \mathcal{M})$ with an equivalence relation $\sim$ such that the quotient map $\mathrm{pr}: M \longrightarrow M/\sim$ is an open map like e.g. for an orbit space with respect to a group action by homeomorphisms, see also Exercise 3.4.8. Show that the quotient $M/\sim$ is locally compact again.

Chapter 6
Continuous Functions

It is one of the main insights in geometry that there is, at least morally, a duality between spaces and the algebras of functions on these spaces: the examples and situations where this idea can be made precise are manifold and we will not go into the details here. Instead, we take this as a motivation to study the algebra of continuous functions on topological spaces more closely.

6.1 Urysohn's Lemma and Tietze's Theorem

The first major question is of course whether or not there are non-trivial continuous functions on a given topological space: we know that for an indiscrete topological space $(M, \mathcal{M}_{\text{indiscrete}})$ there are no other continuous functions $f \colon M \longrightarrow \mathbb{C}$ beside the constant ones. Thus in this situation $\mathscr{C}(M) \simeq \mathbb{C}$ and there is no way of reconstructing M from $\mathscr{C}(M)$ at all. Urysohn's Lemma gives now a remarkable construction of continuous functions for the case of a T_4-space. The basic idea is to squeeze in between two disjoint closed subsets another third subset such that we can get a (still discontinuous) step function. Now this will be iterated and the limit of the step functions yields the continuous function one is looking for:

Theorem 6.1.1 (Urysohn's Lemma) *Let $(M, \mathcal{M})$ be a topological space. Then the following statements are equivalent:*

(i) The space $(M, \mathcal{M})$ is a T_4-space.
(ii) For any non-empty disjoint and closed subsets $A, B \subseteq M$ there exists a continuous function $f \colon M \longrightarrow [0, 1]$ with

$$f\big|_A = 0 \quad \text{and} \quad f\big|_B = 1. \tag{6.1.1}$$

Proof Suppose (ii) holds and f satisfies (6.1.1). Then $\mathcal{O}_1 = f^{-1}([0, \frac{1}{2}))$ and $\mathcal{O}_2 = f^{-1}((\frac{1}{2}, 1])$ are disjoint open subsets with $A \subseteq \mathcal{O}_1$ and $B \subseteq \mathcal{O}_2$. Hence disjoint

© Springer International Publishing Switzerland 2014
S. Waldmann, *Topology*, DOI 10.1007/978-3-319-09680-3_6

Fig. 6.1 Separating A and B
by G_0

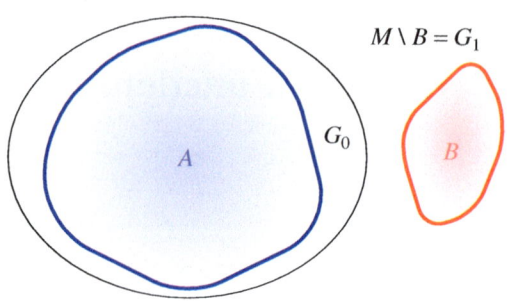

$$M \setminus B = G_1$$

closed subsets can be separated by open subsets showing that $(M, \mathcal{M})$ is T_4. It is the implication (i) $\implies$ (ii) that is non-trivial. To construct f for given A and B we consider the set of positive dyadic numbers, i.e.

$$\mathcal{D} = \{p2^{-q} \mid p, q \in \mathbb{N}\}.$$

For $t \in \mathcal{D}$ with $t > 1$ we set $G_t = M$ and $G_1 = M \setminus B$. Moreover, we choose an open subset $G_0 \subseteq M$ with $A \subseteq G_0 \subseteq G_0^{\mathrm{cl}} \subseteq M \setminus B$ which is possible according to Proposition 2.6.6 since $A \subseteq M \setminus B$ and $M \setminus B$ is open, see Fig. 6.1. The numbers in $\mathcal{D}$ between 0 and 1 can now be enumerated by considering $t = (2m - 1)2^{-n}$ for $n \geq 1$ and $m = 1, \ldots, 2^{n-1}$. This way, every number $t \in \mathcal{D}$ with $0 < t < 1$ determines a unique n and m. We want to construct inductively on n open subsets G_t with $G_t \subseteq G_t^{\mathrm{cl}} \subseteq G_{t'}$ whenever $0 \leq t < t' \leq 1$. For $n = 1$ we only have one $m = 1$ leading to $t = \frac{1}{2}$. Since $G_0^{\mathrm{cl}} \subseteq G_1$ we can squeeze in an open subset $G_{\frac{1}{2}}$ between G_0^{cl} and G_1, i.e. $G_{\frac{1}{2}}^{\mathrm{cl}}$ is still contained in G_1, again by Proposition 2.6.6. Suppose we have constructed G_t for t with denominators $\leq 2^{n-1}$. Then all numbers $k2^{-n+1}$ occurred already for $k = 0, \ldots, 2^{n-1}$ and we have to fill in the gaps in the middle of each consecutive pair. We can find $G_{\frac{2k+1}{2^n}}$ with

$$G^{\mathrm{cl}}_{\frac{k}{2^{n-1}}} \subseteq G_{\frac{2k+1}{2^n}} \subseteq G^{\mathrm{cl}}_{\frac{2k+1}{2^n}} \subseteq G_{\frac{k+1}{2^{n-1}}},$$

again by Proposition 2.6.6, see Fig. 6.2. The desired function f is now defined by

$$f(p) = \inf\{t \mid p \in G_t\}. \tag{$*$}$$

Fig. 6.2 Inserting $G_{\frac{2k+1}{2^n}}$
between $G_{\frac{k}{2^{n-1}}}$ and $G_{\frac{k+1}{2^{n-1}}}$

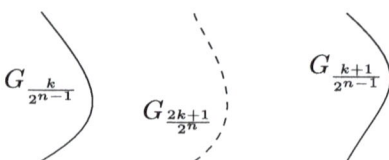

First we note that clearly $f|_A = 0$ and $f|_B = 1$ since $A \subseteq G_0$ and $M \setminus B = G_1$. Moreover, it is clear that f only takes values between 0 and 1. It remains to check the continuity of f. Let $p \in M$ be given and let $\varepsilon > 0$ such that we have an open interval $f(p) \in (f(p) - \varepsilon, f(p) + \varepsilon)$ around the value $f(p)$. Now choose $0 < \delta, \delta' < \varepsilon$ such that $f(p) + \delta, f(p) - \delta' \in \mathcal{D}$. Then for $q \in G_{f(p)+\delta}$ we have $f(q) \leq f(p) + \delta$ and thus $f(q) < f(p) + \varepsilon$. For $q \notin G^{\mathrm{cl}}_{f(p)-\delta'}$ we have $f(q) \geq f(p) - \delta'$ and hence $f(q) > f(p) - \varepsilon$. This shows that for $q \in G_{f(p)+\varepsilon} \setminus G^{\mathrm{cl}}_{f(p)-\varepsilon}$ we get $f(q) \in (f(p) - \varepsilon, f(p) + \varepsilon)$. Since $G_{f(p)+\delta} \setminus G^{\mathrm{cl}}_{f(p)-\delta'}$ is open, it is a neighbourhood of p, which is mapped into the neighbourhood $(f(p) - \varepsilon, f(p) + \varepsilon)$ of $f(p)$. Thus $f^{-1}((f(p) - \varepsilon, f(p) + \varepsilon))$ contains $G_{f(p)+\delta} \setminus G^{\mathrm{cl}}_{f(p)-\delta}$ and is therefore a neighbourhood itself showing the continuity of f at p. $\square$

Remark 6.1.2 We know already many examples of T_4-spaces, e.g. every metric space (Proposition 2.6.9) and every compact Hausdorff space (Proposition 5.1.13). Note also that there is nothing special about the values 0 and 1. We can also find a continuous function $f : M \longrightarrow \mathbb{R}$ such that

$$f|_A = a \quad \text{and} \quad f|_B = b, \tag{6.1.2}$$

where $a, b \in \mathbb{R}$ are some fixed numbers.

A locally compact Hausdorff space is, in general, only regular but not T_4. Nevertheless, for such a space we still have interesting continuous functions:

Corollary 6.1.3 *Let $(M, \mathcal{M})$ be a locally compact Hausdorff space. For a nonempty compact subset $K \subseteq M$ and an open subset $\mathcal{O} \subseteq M$ with $K \subseteq \mathcal{O}$ we find a continuous function $f : M \longrightarrow [0, 1]$ with*

$$f|_K = 0 \quad \text{and} \quad f|_{M \setminus \mathcal{O}} = 1. \tag{6.1.3}$$

Again, we can replace the values 0 and 1 by any other $a, b \in \mathbb{R}$.

Proof Let $B \subseteq M$ be an arbitrary closed subset disjoint from K. Then there is an open subset $U \subseteq M$ with compact closure U^{cl} such that

$$K \subseteq U \subseteq U^{\mathrm{cl}} \subseteq M \setminus B. \tag{$*$}$$

Indeed, let $U_p \in \mathfrak{U}(p)$ be open neighbourhoods of $p \in K$ with compact closure such that $U_p^{\mathrm{cl}} \subseteq M \setminus B$. Since every point has a neighbourhood basis of compact neighbourhoods according to Corollary 5.4.3, this is possible. Then the compactness of K allows to choose finitely many $U_{p_1}, \ldots, U_{p_n}$ with $K \subseteq U = U_{p_1} \cup \cdots \cup U_{p_n}$. Now $U^{\mathrm{cl}} = U_{p_1}^{\mathrm{cl}} \cup \cdots \cup U_{p_n}^{\mathrm{cl}}$ is still compact and contained in $M \setminus B$. This shows $(*)$. Starting now with $A = K$ and $B = M \setminus \mathcal{O}$, we can proceed from here as in the proof of Urysohn's Lemma. $\square$

While Urysohn's Lemma provides us with many non-trivial functions, Tietze's Theorem allows to extend a continuous function defined on a closed subset to the whole space in a continuous way. For its proof we will need the following technical lemma:

Lemma 6.1.4 *Let $(M, \mathcal{M})$ be a T_4-space and let $A \subseteq M$ be a non-empty closed subset. Let $f \in \mathscr{C}_b(A, \mathbb{R})$ be a bounded continuous function on A. Then there exists a function $g \in \mathscr{C}(M, \mathbb{R})$ with*

$$|g(p) - f(p)| \leq \tfrac{2}{3}\|f\|_\infty \quad for \quad p \in A, \tag{6.1.4}$$

and

$$\|g\|_\infty \leq \tfrac{1}{3}\|f\|_\infty, \tag{6.1.5}$$

where $\|\cdot\|_\infty$ denotes the usual supremum norm over the domain where the function is defined, i.e. either A or M.

Proof We decompose A into three pieces where f takes the following values

$$B = f^{-1}\left(\left[-\|f\|_\infty, -\tfrac{1}{3}\|f\|_\infty\right]\right), \quad C = f^{-1}\left(\left[\tfrac{1}{3}\|f\|_\infty, \|f\|_\infty\right]\right),$$

and the remaining part where $-\tfrac{1}{3}\|f\|_\infty < f < \tfrac{1}{3}\|f\|_\infty$. Clearly, $B \cap C = \emptyset$ and $B, C \subseteq A$ are closed in A since f is continuous. Since A is closed, $B, C \subseteq M$ are still closed in M. We set $g|_B = -\tfrac{1}{3}\|f\|_\infty$ and $g|_C = \tfrac{1}{3}\|f\|_\infty$. Then g is continuous on $B \cup C$ and takes values between $-\tfrac{1}{3}\|f\|_\infty$ and $\tfrac{1}{3}\|f\|_\infty$. By Urysohn's Lemma we can extend g continuously to M such that we still have values between $-\tfrac{1}{3}\|f\|_\infty$ and $\tfrac{1}{3}\|f\|_\infty$. Hence $\|g\|_\infty \leq \tfrac{1}{3}\|f\|_\infty$. Moreover, on B we have $|f(p) - g(p)| \leq \tfrac{2}{3}\|f\|_\infty$ for all $p \in B$ since f takes values between $-\|f\|_\infty$ and $-\tfrac{1}{3}\|f\|_\infty$. Analogously we have $|f(p) - g(p)| \leq \tfrac{2}{3}\|f\|_\infty$ for $p \in C$. Finally, for $p \in A \setminus (B \cup C)$ we have $|f(p)| < \tfrac{1}{3}\|f\|_\infty$ and hence also here $|f(p) - g(p)| \leq \tfrac{2}{3}\|f\|_\infty$, resulting in (6.1.4). $\qquad\square$

Theorem 6.1.5 (Tietze) *Let $(M, \mathcal{M})$ be a T_4-space and let $A \subseteq M$ be closed. If $f: A \longrightarrow [a, b]$ is a continuous function on A then f has a continuous extension $F: M \longrightarrow [a, b]$.*

Proof By adding a constant and thus a continuous function we can assume that $a = -\|f\|_\infty$ and $b = \|f\|_\infty$. Now let $g_1: M \longrightarrow [-\tfrac{1}{3}\|f\|_\infty, \tfrac{1}{3}\|f\|_\infty]$ be a function as in Lemma 6.1.4 and consider $f_2 = f - g_1$. This is a continuous function on A with $\|f_2\|_\infty \leq \tfrac{2}{3}\|f\|_\infty$. Thus we find a continuous function $g_2: M \longrightarrow [-\tfrac{1}{3}\|f_1\|_\infty, -\tfrac{1}{3}\|f_1\|_\infty] = [-\tfrac{1}{3}\tfrac{2}{3}\|f\|_\infty, \tfrac{1}{3}\tfrac{2}{3}\|f\|_\infty]$ with $\|f_2 - g_2\|_\infty \leq \tfrac{2}{3}\|f_2\|_\infty \leq (\tfrac{2}{3})^2\|f\|_\infty$ and thus

$$\|f - g_1 - g_2\|_\infty \leq \left(\tfrac{2}{3}\right)^2 \|f\|_\infty.$$

Inductively, this gives continuous functions $g_n: M \longrightarrow \mathbb{R}$ with

$$\|g_n\|_\infty \le \tfrac{1}{3} \left(\tfrac{2}{3}\right)^{n-1} \|f\|_\infty \qquad (*)$$

and on A

$$\|f - g_1 - \cdots - g_n\|_\infty \le \left(\tfrac{2}{3}\right)^n \|f\|_\infty. \qquad (**)$$

Now on M the series $\sum_{n=1}^\infty g_n$ converges absolutely with respect to the sup-norm to a function g with

$$\|g\|_\infty \le \sum_{n=1}^\infty \|g_n\|_\infty \overset{(*)}{\le} \sum_{n=1}^\infty \tfrac{1}{3} \left(\tfrac{2}{3}\right)^{n-1} \|f\|_\infty = \tfrac{1}{3} \cdot \frac{\|f\|_\infty}{1 - \tfrac{2}{3}} = \|f\|_\infty.$$

Since the convergence is uniform thanks to $\|\cdot\|_\infty$, the limit g is a continuous function again, see Exercise 2.7.19, (vii). Moreover, for $p \in A$ we have by

$$f(p) = g(p).$$

In fact, the convergence $f = \sum_{n=1}^\infty g_n\big|_A$ is even uniform, Hence g is the desired extension of f. $\qquad \square$

This extension theorem has several variants also dealing with unbounded continuous functions or vector-valued functions, see e.g. the discussion in [13, Sect. 8.3]. Moreover, the Tietze Extension Theorem has a reverse statement characterizing T_4-spaces:

Proposition 6.1.6 *Let $(M, \mathcal{M})$ be a topological space and assume that for every closed subset $A \subseteq M$ and every continuous function $f : A \longrightarrow [-1, 1]$ there is a continuous extension $F : M \longrightarrow \mathbb{R}$. Then $(M, \mathcal{M})$ is a T_4-space.*

Proof Assume $(M, \mathcal{M})$ has this property and let $A, B \subseteq M$ be closed disjoint subsets. Then on the closed subset $C = A \cup B$ we define $f : C \longrightarrow [-1, 1]$ by

$$f\big|_A = -1 \quad \text{and} \quad f\big|_B = 1.$$

Since $A \cap B = \emptyset$ this gives a continuous function having a continuous extension F by assumption. Define

$$\mathcal{O}_1 = F^{-1}\left(\left(-\infty, -\tfrac{1}{2}\right)\right) \quad \text{and} \quad \mathcal{O}_2 = F^{-1}\left(\left(\tfrac{1}{2}, \infty\right)\right),$$

which are open subsets by the continuity of F. Clearly $\mathcal{O}_1 \cap \mathcal{O}_2 = \emptyset$ and $A \subseteq \mathcal{O}_1$ and $B \subseteq \mathcal{O}_2$. Thus we can separate A and B, showing that M is indeed T_4. $\qquad \square$

6.2 The Stone-Weierstraß Theorem

The classical theorem of Weierstraß says that a continuous function on [0, 1] can be approximated uniformly by polynomials. There is even a rather explicit construction using the *Bernstein polynomials*, see e.g. [2, p. 323]. This statement has a drastic generalization which on one hand uncovers the core feature of polynomials needed, and, on the other hand, makes it one of the most important tools in large areas of functional analysis. We consider the complex-valued continuous functions $\mathscr{C}(M)$ on a compact Hausdorff space $(M, \mathcal{M})$. With M being compact, we know that a continuous function is bounded and thus the supremum norm is well-defined on $\mathscr{C}(M) = \mathscr{C}_b(M)$. From Exercise 2.7.19 we know that $\mathscr{C}(M)$ is a complex Banach space with the additional structure of an associative commutative algebra, using the pointwise product of functions. Finally, we can take the complex conjugation as a *-involution: it is an anti-linear, involutive anti-automorphism of the algebra multiplication. The compatibility between the product, the involution, and the supremum norm can be summarized as follows:

Definition 6.2.1 (*C^*-Algebra*) Let $\mathcal{A}$ be a complex associative algebra with $*$ being a *-involution and $\| \cdot \|$ being a norm on $\mathcal{A}$.

(i) $(\mathcal{A}, \| \cdot \|)$ is called a normed algebra if for all $a, b \in \mathcal{A}$ one has

$$\|ab\| \leq \|a\|\|b\|. \tag{6.2.1}$$

(ii) $(\mathcal{A}, \| \cdot \|, {}^*)$ is called a normed *-algebra if in addition

$$\|a^*\| = \|a\|. \tag{6.2.2}$$

(iii) $(\mathcal{A}, \| \cdot \|)$ is called a Banach algebra, and $(A, \| \cdot \|, {}^*)$ is called a Banach *-algebra, respectively, if in addition the norm is complete.

(iv) $(\mathcal{A}, \| \cdot \|, {}^*)$ is called a C^*-algebra if it is a Banach *-algebra with

$$\|a^*a\| = \|a\|^2. \tag{6.2.3}$$

The previous results of Exercise 2.7.19 can now be summarized as follows:

Proposition 6.2.2 *Let $(M, \mathcal{M})$ be a compact Hausdorff space. Then $\mathscr{C}(M) = \mathscr{C}_b(M)$ is a commutative C^*-Algebra with unit element $\mathbb{1}$ being the function constant 1.*

Remark 6.2.3 Needless to say, there is a sophisticated and beautiful theory of C^*-algebras which can be found e.g. in the classic textbooks [1,7,15,16,30] for a first reading. Without going into any details we just mention that $\mathscr{C}(M)$ is *the* commutative unital C^*-algebra: every commutative unital C^*-algebra is isomorphic to some $\mathscr{C}(M)$ with a compact Hausdorff space M being determined uniquely up to homeomorphism. This is the core result of the Gel'fand-Naimark theory which is heavily based on the Banach-Alaoglu Theorem, a consequence of Tikhonov's Theorem, as we already mentioned.

We are now interested in the following situation: given M one often has a nicer class of functions inside $\mathscr{C}(M)$ like the polynomials on $[0, 1]$ or smooth functions if M happens to be a smooth manifold etc. In this case many constructions are first done for this nicer class of functions. In a second step one wants to extend the achieved results to the larger class $\mathscr{C}(M)$. Of course, in general such an extension will cause problems: e.g. there is no obvious way to extend differentiation from differentiable functions to continuous ones. Nevertheless, in many situations one still has some justified hope that this program is reasonable. The remaining question is whether or not the nice class of functions is already "large enough" to pass to all continuous functions. Here we have a very easy way to make this precise: since $\mathscr{C}(M)$ is a Banach space, it carries a topology which is even a metric topology. Thus we can ask whether or not a given subset $\mathcal{A} \subseteq \mathscr{C}(M)$ is dense, i.e. $\mathcal{A}^{\mathrm{cl}} = \mathscr{C}(M)$. This is precisely the situation dealt with in the Stone-Weierstraß Theorem. We consider a $*$-subalgebra $\mathcal{A} \subseteq \mathscr{C}(M)$, i.e. a subspace which is closed with respect to products and the $*$-involution. Then the following proposition is straightforward:

Proposition 6.2.4 *Let $(M, \mathcal{M})$ be a compact Hausdorff space and let $\mathcal{A} \subseteq \mathscr{C}(M)$ be a $*$-subalgebra. Then $\mathcal{A}^{\mathrm{cl}}$ is a C^{*}-subalgebra of $\mathscr{C}(M)$.*

Proof The first observation is that all algebraic operation are continuous, i.e. the maps

$$\mathscr{C}(M) \times \mathscr{C}(M) \ni (f, g) \mapsto f + g \in \mathscr{C}(M),$$
$$\mathbb{C} \times \mathscr{C}(M) \ni (z, f) \mapsto zf \in \mathscr{C}(M),$$
$$\mathscr{C}(M) \times \mathscr{C}(M) \ni (f, g) \mapsto fg \in \mathscr{C}(M),$$
$$\mathscr{C}(M) \ni f \mapsto \overline{f} \in \mathscr{C}(M)$$

are continuous when we take the product topologies on the left sides. This is a general fact in a normed $*$-algebra and does not yet refer to the particular structure of $\mathscr{C}(M)$, see Exercise 6.4.2. Using this continuity, we can argue with sequences f_n approximating a function $f \in \mathcal{A}^{\mathrm{cl}}$ in the closure since we are in a metric and hence first countable space, see Proposition 4.1.9, (ii). Thus let $f, g \in \mathcal{A}^{\mathrm{cl}}$ and choose sequences $f_n, g_n \in \mathcal{A}$ with $f_n \longrightarrow f$ and $g_n \longrightarrow g$. Then $f_n + g_n \longrightarrow f + g$, $zf_n \longrightarrow zf$, $f_n g_n \longrightarrow fg$, and $\overline{f_n} \longrightarrow \overline{f}$ follows from the continuity of the above structure maps. Again, this holds in every normed $*$-algebra and thus the closure of a $*$-subalgebra is again a $*$-subalgebra. The last step is to show that $\mathcal{A}^{\mathrm{cl}}$ is a C^{*}-algebra itself. First we note that the C^{*}-property (6.2.3) is of course still valid as it holds for all functions in $\mathscr{C}(M)$. It remains to check the completeness: let $f_n \in \mathcal{A}^{\mathrm{cl}}$ be a Cauchy sequence in $\mathcal{A}^{\mathrm{cl}}$ then the completeness of $\mathscr{C}(M)$ provides us a limit $f \in \mathscr{C}(M)$ with $f_n \longrightarrow f$. But then $f \in \mathcal{A}^{\mathrm{cl}}$ by Proposition 4.1.7. Thus $\mathcal{A}^{\mathrm{cl}}$ is complete as well. This argument works in a general metric space, see also Exercise 4.4.1. $\square$

Since we can approximate the square root uniformly on $[0, 1]$ by polynomials p_n according to Exercise 6.4.1, we get the following approximation result for C^*-subalgebras of $\mathscr{C}(M)$:

Proposition 6.2.5 *Let $(M, \mathcal{M})$ be a compact Hausdorff space and let $\mathcal{A} \subseteq \mathscr{C}(M)$ be a C^*-subalgebra.*

   (i) If $f \in \mathcal{A}$ with $f \geq 0$ then $\sqrt{f} \in \mathcal{A}$.
   (ii) If $f \in \mathcal{A}$ then $|f| \in \mathcal{A}$.
   (iii) If $f, g \in \mathcal{A}$ are real-valued then $\min(f, g), \max(f, g) \in \mathcal{A}$.

Proof Let $f \in \mathcal{A}$ with $f \geq 0$ be given. Then $f(p) \in [0, \|f\|_\infty]$ for all $p \in M$ by Proposition 5.2.6. Without restriction, we can assume $\|f\|_\infty \neq 0$ since otherwise $f = 0$ clearly has a square root in $\mathcal{A}$. Consider now the continuous function $p_n \left(\frac{f}{\|f\|_\infty} \right) : M \longrightarrow [0, 1]$ where p_n is given as in Exercise 6.4.1. Since $p_n(0) = 0$ the polynomial p_n has no constant term. Thus the function $p_n \left(\frac{f}{\|f\|_\infty} \right)$ is again in the subalgebra $\mathcal{A}$ since $\mathcal{A}$ being a subalgebra is closed under the algebraic operations of pointwise addition and multiplication of functions. Moreover, for all $p \in M$ we have

$$\left| p_n \left(\frac{f(p)}{\|f\|_\infty} \right) - \sqrt{\frac{f(p)}{\|f\|_\infty}} \right| \leq \frac{2 \sqrt{\frac{f(p)}{\|f\|_\infty}}}{2 + n \sqrt{\frac{f(p)}{\|f\|_\infty}}} \leq \frac{2}{n},$$

and thus

$$\left\| p_n \left(\frac{f}{\|f\|_\infty} \right) - \sqrt{\frac{f}{\|f\|_\infty}} \right\|_\infty \longrightarrow 0$$

for $n \longrightarrow \infty$. Hence the functions $\sqrt{\|f\|_\infty}\, p_n \left(\frac{f}{\|f\|_\infty} \right) \in \mathcal{A}$ approximate $\sqrt{f}$ uniformly. Since by assumption $\mathcal{A} = \mathcal{A}^{\mathrm{cl}}$ is closed, we have $\sqrt{f} \in \mathcal{A}$, showing the first part. For $f \in \mathcal{A}$ we have $\overline{f} f \in \mathcal{A}$ since $\mathcal{A}$ is a *-algebra. Then the first part shows $|f| = \sqrt{\overline{f} f} \in \mathcal{A}$, too. The last part is clear since

$$\min(f, g) = \tfrac{1}{2}(f + g - |f - g|) \quad \text{and} \quad \max(f, g) = \tfrac{1}{2}(f + g + |f - g|),$$

and all functions on the right belong to $\mathcal{A}$.   □

For a *-subalgebra $\mathcal{A} \subseteq \mathscr{C}(M)$ the interesting question is now how large the corresponding C^*-subalgebra $\mathcal{A}^{\mathrm{cl}} \subseteq \mathscr{C}(M)$ can be. In particular, we are interested in the situation where $\mathcal{A}^{\mathrm{cl}} = \mathscr{C}(M)$, i.e. $\mathcal{A}$ is *dense* in $\mathscr{C}(M)$. The theorem of Stone-Weierstraß gives a simple criterion for this:

Theorem 6.2.6 (Stone-Weierstraß) *Let $(M, \mathcal{M})$ be a compact Hausdorff space and let $\mathcal{A} \subseteq \mathscr{C}(M)$ be a point-separating unital *-subalgebra. Then $\mathcal{A}^{\mathrm{cl}} = \mathscr{C}(M)$.*

Here *point-separating* means that for different points $p \neq q$ we can find a function $f \in \mathcal{A}$ with $f(p) \neq f(q)$. Inside $\mathscr{C}(M)$ we know to have such functions thanks to Urysohn's Lemma. *Unital* means that the unit $\mathbb{1} \in \mathscr{C}(M)$ is contained in $\mathcal{A}$.

Proof (of Theorem 6.2.6) We have to show that for $f \in \mathscr{C}(M)$ and $\epsilon > 0$ we find a $g_\epsilon \in \mathcal{A}$ with $\|f - g_\epsilon\|_\infty < \epsilon$. Since $\mathcal{A}$ is dense in $\mathcal{A}^{\mathrm{cl}}$, it suffices to find such a function g_ϵ in $\mathcal{A}^{\mathrm{cl}}$ by the triangle inequality for $\| \cdot \|_\infty$. Since $f = \mathrm{Re}(f) + i\mathrm{Im}(f)$ and since $\mathcal{A}$ is assumed to be a *-algebra, we can approximate real and imaginary parts separately. Hence we can assume $f = \overline{f}$ from the beginning. Since $\mathcal{A}$ is a *-algebra we get for $p \neq q$ a function $g = \overline{g} \in \mathcal{A}$ with $g(p) \neq g(q)$. For this g we consider

$$h = \frac{f(p) - f(q)}{g(p) - g(q)} g - \frac{f(p)g(q) - f(q)g(p)}{g(p) - g(q)} \mathbb{1}.$$

Since $\mathbb{1} \in \mathcal{A}$ we get $h = \overline{h} \in \mathcal{A}$. Moreover, we have

$$h(p) = f(p) \quad \text{and} \quad h(q) = f(q). \tag{$*$}$$

If $p = q$ we take the constant function $h = f(p)\mathbb{1}$. This way, we get for any two points $p, q \in M$ a function $h \in \mathcal{A}$ with $(*)$. In a next step we fix a point $q \in M$. Then for every $p \in M$ we find a function $h_p \in \mathcal{A}$ with $(*)$. The subset

$$U_p = \{p' \in M \mid h_p(p') < f(p') + \epsilon\}$$

is open as h_p and f are continuous. We have $p \in U_p$. This way we get an open cover $\{U_p\}_{p \in M}$ of M. Let $p_1, \ldots, p_n \in M$ be finitely many points such that $M = U_{p_1} \cup \cdots \cup U_{p_n}$ as M is compact. We define now the continuous function

$$g_q = \min\{h_{p_1}, \ldots, h_{p_n}\}.$$

For g_q we have for all $p \in M$ the estimate

$$g_q(p) < f(p) + \epsilon, \tag{$**$}$$

and of course $g_q(q) = f(q)$. According to Proposition 6.2.5, (iii), we have $g_q \in \mathcal{A}^{\mathrm{cl}}$. Now

$$V_q = \{p \in M \mid g_q(p) > f(p) - \epsilon\}$$

is again open and $q \in V_q$. Thus we have an open cover $\{V_q\}_{q \in M}$ of M with a finite subcover $V_{q_1}, \ldots, V_{q_m}$. We finally set

$$g_\epsilon = \max\{g_{q_1}, \ldots, g_{q_m}\},$$

which is again a function in $\mathcal{A}^{\mathrm{cl}}$ by Proposition 6.2.5, (iii). Moreover, for every $q, p \in M$ we have (∗∗) and hence $g_\epsilon(p) < f(p) + \epsilon$ by the very definition of g_ϵ and the choice of the $V_{q_1}, \ldots, V_{q_m}$. Thus $\|g_\epsilon - f\|_\infty < \epsilon$, proving the theorem. $\square$

Remark 6.2.7 The Stone-Weierstraß Theorem has many variations and alternative formulations, some of which are discussed in Exercise 6.4.4. In order to appreciate this result, one should remember that continuous functions, even just on the compact interval $[0, 1]$, can behave rather wildly, even though polynomials can approximate them uniformly, see Exercise 6.4.7.

We shall now formulate a version of the Stone-Weierstraß Theorem for locally-compact Hausdorff-spaces. The basic idea is that we can not expect to achieve a uniform approximation in this case but at least a *locally* uniform one: the used topology to approximate continuous functions is the locally convex topology arising from all the seminorms

$$\|f\|_K = \sup_{p \in K} |f(p)|, \tag{6.2.4}$$

where $K \subseteq M$ is compact. The resulting topology for $\mathscr{C}(M)$ is discussed in Exercise 6.4.3. The following statement is then an easy transfer of Theorem 6.2.6 to the locally compact framework:

Theorem 6.2.8 *Let $(M, \mathcal{M})$ be a locally compact Hausdorff space and let $\mathcal{A} \subseteq \mathscr{C}(M)$ be a point-separating unital ∗-subalgebra. Then $\mathcal{A}^{\mathrm{cl}} = \mathscr{C}(M)$ with respect to locally uniform convergence.*

Proof The proof is now very simple: for $f \in \mathscr{C}(M)$ and $\epsilon > 0$ and $K \subseteq M$ compact, we have to show that there is a function $g_{\epsilon,K} \in \mathcal{A}$ with $\|f - g_{\epsilon,K}\|_K < \epsilon$. Now

$$\mathcal{A}_K = \left\{ g\big|_K \in \mathscr{C}(K) \mid g \in \mathcal{A} \right\} \subseteq \mathscr{C}(K)$$

is clearly a ∗-subalgebra of $\mathscr{C}(K)$ which is still unital and point-separating. Thus there is a $g_\epsilon \in \mathcal{A}_K$ with $\|f\big|_K - g_\epsilon\|_K < \epsilon$ by Theorem 6.2.6. For this g_ϵ we have a $g_{\epsilon,K} \in \mathcal{A}$ with $g_\epsilon = g_{\epsilon,K}\big|_K$ by the very definition of $\mathcal{A}_K$. Then $\|f - g_{\epsilon,K}\|_K = \|f\big|_K - g_\epsilon\|_K < \epsilon$ as wanted. $\square$

Note that the topology of locally uniform convergence is typically no longer first countable and hence we only get a net $\{g_i\}_{i \in I}$ in $\mathcal{A}$ converging to $f \in \mathcal{A}^{\mathrm{cl}}$ but not a sequence in general. The above construction gives only a net with index set $\{(\epsilon, K) \mid \epsilon > 0, K \subseteq M \text{ compact}\}$ endowed with its obvious direction.

The following class of locally compact spaces gives a first countable topology for $\mathscr{C}(M)$:

Definition 6.2.9 (σ-*compactness*) A topological space $(M, \mathcal{M})$ is called σ-compact if there is a sequence of compact subsets $\{K_n\}_{n \in \mathbb{N}}$ such that

$$M = \bigcup_{n=1}^{\infty} K_n. \tag{6.2.5}$$

Proposition 6.2.10 *Let $(M, \mathcal{M})$ be a locally compact Hausdorff space which is second countable. Then we have:*

(i) *There exists a countable basis $\{\mathcal{O}_n\}_{n\in\mathbb{N}}$ of the topology with compact closures $\mathcal{O}_n^{\mathrm{cl}}$.*

(ii) *There exists a sequence of compact subsets $\{K_n\}_{n\in\mathbb{N}}$ such that for all $n \in \mathbb{N}$ one has $K_n \subseteq K_{n+1}^\circ$ and*

$$M = \bigcup_{n=1}^\infty K_n. \tag{6.2.6}$$

In particular, M is σ-compact.

(iii) *The topology of locally uniform convergence for $\mathscr{C}(M)$ is first countable.*

Proof Let $\{\mathcal{O}_n\}_{n\in\mathbb{N}}$ be a countable basis of the topology $\mathcal{M}$ of M. From Corollary 5.4.3 we know that there are neighbourhood bases for every point $p \in M$ consisting of compact subsets. Thus there are neighbourhood bases consisting of open subsets with compact closures as well, by taking the open interiors K° of the compact neighbourhoods K. Since every such K° is a union of certain $\mathcal{O}_n$'s we see that among the $\mathcal{O}_n$'s we have open subsets with compact closures: this is the case as soon as $\mathcal{O}_n \subseteq K^\circ$ for some compact K. Since the K° with $p \in K$ and K compact provide a neighbourhood basis of each $p \in M$, the collection of all the $\{K^\circ\}_{K \text{ compact}}$ provides a basis of the topology of M. But then it is clear that the $\mathcal{O}_n$'s with $\mathcal{O}_n \subseteq K^\circ$ for some compact K still will be an at most countable basis of $\mathcal{M}$, proving the first part. For the second part, assume that we have a countable basis $\{\mathcal{O}_n\}_{n\in\mathbb{N}}$ of open sets with compact closure $\mathcal{O}_n^{\mathrm{cl}}$. Now define the compact subset $K_1 = \mathcal{O}_1^{\mathrm{cl}}$. By compactness of K_1 we find finitely many $\mathcal{O}_{n_1}, \ldots, \mathcal{O}_{n_k}$ which cover K_1. Thus we set $K_2 = (\mathcal{O}_2 \cup \mathcal{O}_{n_1} \cup \cdots \cup \mathcal{O}_{n_k})^{\mathrm{cl}} = \mathcal{O}_2^{\mathrm{cl}} \cup \mathcal{O}_{n_1}^{\mathrm{cl}} \cup \cdots \cup \mathcal{O}_{n_k}^{\mathrm{cl}}$, which is again compact and satisfies $K_1 \subseteq K_2^\circ$. Proceeding inductively will give us the sequence of exhausting compact subsets as wanted. For the third part, let $K \subseteq M$ be compact. Since the $\{K_n^\circ\}_{n\in\mathbb{N}}$ provide an open cover with $K_n^\circ \subseteq K_{n+1}^\circ$ there is a $n \in \mathbb{N}$ with $K \subseteq K_{n-1}^\circ \subseteq K_n$. Clearly $\|f\|_K \le \|f\|_{K_n}$ for all $f \in \mathscr{C}(M)$. This shows that the open ϵ-ball around $f \in \mathscr{C}(M)$ with respect to the seminorm $\|\cdot\|_K$ contains an open ϵ-ball of f with respect to $\|\cdot\|_{K_n}$ since $\|f - g\|_{K_n} < \epsilon$ implies $\|f - g\|_K < \epsilon$. From this, one sees that the ϵ-balls with respect to the seminorms $\|\cdot\|_{K_n}$ provide a neighbourhood basis of f. Finally, it suffices to take any zero sequence $\epsilon_m \longrightarrow 0$ such that we get a countable neighbourhood basis by the open balls $\left\{\mathrm{B}_{\|\cdot\|_{K_n}, \epsilon_m}(f)\right\}_{n,m\in\mathbb{N}}$ around $f \in \mathscr{C}(M)$ with respect to the seminorms $\|\cdot\|_{K_n}$, see also Exercise 6.4.3, and thus the third statement follows. $\qquad\square$

Corollary 6.2.11 *Let $(M, \mathcal{M})$ be a locally compact, second countable Hausdorff space and let $\mathcal{A} \subseteq \mathscr{C}(M)$ be a point-separating unital $*$-subalgebra. Then for every $f \in \mathscr{C}(M)$ there exists a sequence $f_n \in \mathcal{A}$ with $f_n \longrightarrow f$ locally uniformly.*

Example 6.2.12 Clearly $\mathbb{R}^n$ is locally compact, Hausdorff, and second countable. More generally, every topological manifold has this property and hence Corollary 6.2.11 applies. This is one of the reasons to ask for second countability of topological manifolds.

Having the above construction in mind it is a good occasion to introduce the concept of paracompactness: this is not just yet another notion of compactness as we have seen now so many. Instead, it is designed to simplify given open covers in a very useful way:

Definition 6.2.13 (*Paracompactness*) A topological space $(M, \mathcal{M})$ is called paracompact if every open cover $\{\mathcal{O}_i\}_{i \in I}$ has an open refinement $\{U_j\}_{j \in J}$, i.e. an open cover with a map $J \ni j \mapsto i(j) \in I$ such that $U_j \subseteq \mathcal{O}_{i(j)}$, such that the refined cover is locally finite, i.e. for every $p \in M$ there exists an open neighbourhood U with $U \cap U_j = \emptyset$ except for finitely many indices $j \in J$.

The significance is that we can simplify arbitrary open covers to locally finite ones: this is useful for many purposes as we can e.g. add up continuous functions if their supports are locally finite in the following way:

Example 6.2.14 Let $\{f_i\}_{i \in I}$ be a arbitrary collection of continuous functions $f_i \in \mathscr{C}(M)$ such that their supports are locally finite. Then the pointwise definition

$$f(p) = \sum_{i \in I} f_i(p) \tag{6.2.7}$$

yields a well-defined and continuous function $f \in \mathscr{C}(M)$. Indeed, there is an open neighbourhood $U \subseteq M$ of p such that supp $f_i \cap U = \emptyset$ for all but finitely many indices, say $i_1, \ldots, i_n$. Hence the restriction

$$f\big|_U = f_{i_1}\big|_U + \cdots + f_{i_n}\big|_U \tag{6.2.8}$$

is a finite sum of continuous functions and hence continuous itself. As this can be done for every point, $f \in \mathscr{C}(M)$ follows, see also Exercise 6.4.5.

The above construction of the exhausting sequence of compact subsets allows for yet another nice feature of second countable locally compact Hausdorff spaces:

Proposition 6.2.15 *Let $(M, \mathcal{M})$ be a second countable locally compact Hausdorff space. Then M is paracompact.*

Proof Let $\{\mathcal{O}_i\}_{i \in I}$ be an arbitrary open cover of M. Moreover, choose an exhausting sequence $\cdots K_n^\circ \subseteq K_n \subseteq K_{n+1}^\circ \subseteq \cdots$ of compact subsets K_n with $M = \bigcup K_n$ as in Proposition 6.2.10, (ii). Consider the open rings $U_n = K_{n+1}^\circ \setminus K_{n-1}$ with $U_0 = K_1^\circ$. They still cover M, see also Fig. 6.3. Moreover, this cover is now locally finite, in fact, we have $U_{n+1} \cap U_{n-1} = \emptyset$ for all n. Now $U_n^{\mathrm{cl}} \subseteq K_{n+1}^{\mathrm{cl}}$ and hence the closed rings U_n^{cl} are compact. This gives for every n finitely many $i_{n,1}, \ldots, i_{n,m} \in I$ such that $U_n^{\mathrm{cl}} \subseteq \mathcal{O}_{i_{n,1}} \cup \cdots \cup \mathcal{O}_{i_{n,m}}$. Thus we consider the open subsets $W_{n,k} = \mathcal{O}_{i_{n,k}} \cap U_n$. We have $W_{n-1,k} \cap W_{n+1,\ell} = \emptyset$ for all n and all k, ℓ, showing that the $W_{n,k}$ are locally finite, too. For fixed n they cover U_n and hence all the $W_{n,k}$ cover M. Finally, $W_{n,k} \subseteq \mathcal{O}_{i_{n,k}}$ and hence we have found the locally finite refinement. Note that the refinement is even countable and each open subset $W_{n,k}$ of it has compact closure. $\qquad \square$

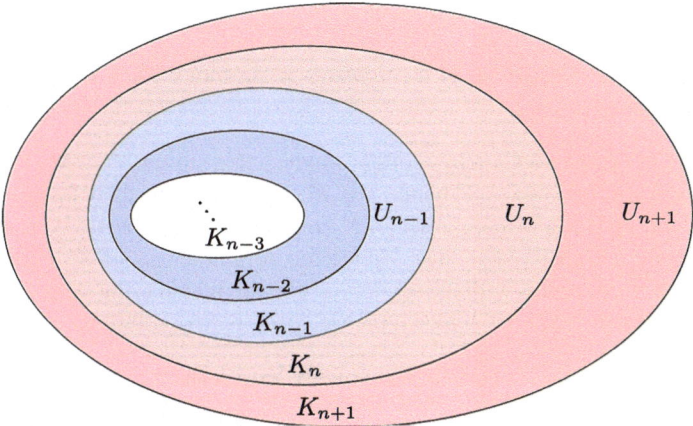

Fig. 6.3 The onion-shaped cover by the U_n. One has $U_{n+1} \cap U_{n-1} = \emptyset$

In particular, open subsets of $\mathbb{R}^n$ as well as topological manifolds are paracompact. This fact is used very much in form of the construction of Example 6.2.14 in various ways in differential geometry.

6.3 The Arzelà-Ascoli Theorem

The Theorem of Arzelà -Ascoli determines the compact subsets of the Banach space $\mathscr{C}(M)$ where $(M, \mathcal{M})$ is a compact Hausdorff space. Since a Banach space is a metric space the notions of compactness, countable compactness, and sequential compactness coincide according to Proposition 5.4.12. Moreover, in a metric space a subset K is compact iff K is complete and totally bounded: recall that a subset K in a metric space is *totally bounded* (sometimes also called *pre-compact*) if for every $\epsilon > 0$ there are finitely many points $p_1, \ldots, p_n \in M$ with

$$K \subseteq \mathrm{B}_\epsilon(p_1) \cup \cdots \cup \mathrm{B}_\epsilon(p_n), \tag{6.3.1}$$

see Exercise 5.5.7 for a detailed discussion on the equivalence of compactness with completeness and total boundedness. Note that we can even assume without restriction that the points $p_1, \ldots, p_n$ are in K: indeed, let $\epsilon > 0$ be given and find $p_1, \ldots, p_n \in M$ with (6.3.1) for $\frac{\epsilon}{2}$ instead of ϵ. Then either $K \cap \mathrm{B}_{\frac{\epsilon}{2}}(p_i) = \emptyset$, in which case we do not need this ball at all, or $K \cap \mathrm{B}_{\frac{\epsilon}{2}}(p_i) \neq \emptyset$. Then choose $q_i \in K \cap \mathrm{B}_{\frac{\epsilon}{2}}(p_i)$. Thanks to the triangle inequality we get $\mathrm{B}_{\frac{\epsilon}{2}}(p_i) \subseteq \mathrm{B}_\epsilon(q_i)$ and thus

$$K \subseteq \mathrm{B}_\epsilon(q_1) \cup \cdots \cup \mathrm{B}_\epsilon(q_n), \tag{6.3.2}$$

follows, now for points $q_1, \ldots, q_n \in K$.

Proposition 6.3.1 *Let (M, d) be a complete metric space. Then $K \subseteq M$ is complete iff K is closed.*

Proof Let $(p_n)_{N \in \mathbb{N}}$ be a Cauchy sequence in a closed subset $K \subseteq M$. Then $p_n \longrightarrow p \in M$ since M is complete and hence $p \in K$ since K is closed, by Proposition 4.1.7. Conversely, if $(p_n)_{n \in \mathbb{N}}$ is a sequence in K converging to some $p \in M$ then $(p_n)_{n \in \mathbb{N}}$ is in particular a Cauchy sequence. Hence the completeness of K implies $p \in K$. Thus $K = K^{\text{scl}}$ coincides with the sequential closure which is the same as the closure K^{cl} by Proposition 4.1.9, (ii), since a metric space is first countable.   □

Also totally bounded subsets behave nicely with respect to closures:

Proposition 6.3.2 *Let (M, d) be a metric space and let $K \subseteq M$. Then K is totally bounded iff K^{cl} is totally bounded.*

Proof First it is clear that any subset of a totally bounded subset is again totally bounded. Hence we only have to show that K^{cl} is totally bounded for a totally bounded K. Thus let $\epsilon > 0$ be given. Then we find $p_1, \ldots, p_n \in K$ with

$$K \subseteq B_{\frac{\epsilon}{2}}(p_1) \cup \cdots \cup B_{\frac{\epsilon}{2}}(p_n),$$

and hence

$$K^{\text{cl}} \subseteq B_{\frac{\epsilon}{2}}(p_1)^{\text{cl}} \cup \cdots \cup B_{\frac{\epsilon}{2}}(p_n)^{\text{cl}}.$$

Since $B_{\frac{\epsilon}{2}}(p_1)^{\text{cl}} \subseteq B_{\epsilon}(p_1)$, the closure K^{cl} is totally bounded again.   □

We consider now a subset $\mathcal{F} \subseteq \mathscr{C}(M)$ of the complex-valued continuous functions on a topological space $(M, \mathcal{M})$. The following definition of equicontinuity will be central in the Arzelà-Ascoli-Theorem:

Definition 6.3.3 (*Equicontinuity*) Let $(M, \mathcal{M})$ be a topological space and let $\mathcal{F} \subseteq \mathscr{C}(M)$ be a subset. Then $\mathcal{F}$ is called equicontinuous at $p \in M$ if for all $\epsilon > 0$ there exists a neighbourhood $U_p \in \mathfrak{U}(p)$ of p such that for all $f \in \mathcal{F}$ and $q \in U_p$ one has

$$|f(p) - f(q)| < \epsilon. \tag{6.3.3}$$

The set $\mathcal{F}$ is called equicontinuous if it is equicontinuous at all points.

If $\mathcal{F} = \{f\}$ consists of a single function then equicontinuity just means that f is continuous at every $p \in M$, i.e. f is continuous. Analogously, for finitely many functions $\mathcal{F} = \{f_1, \ldots, f_n\}$ the continuity at $p \in M$ gives us for $\epsilon > 0$ neighbourhoods $U_q^{(1)}, \ldots, U_q^{(n)} \in \mathfrak{U}(p)$ such that

$$|f_i(p) - f_i(q)| < \epsilon \tag{6.3.4}$$

for $i = 1, \ldots, n$ and $q \in U_p^{(i)}$. Taking $U_p = U_q^{(1)} \cap \cdots \cap U_q^{(n)}$ shows that also in this case, $\mathcal{F}$ is equicontinuous at p. Thus the idea is to control the "degree of continuity" for an *infinite* set of continuous functions and find one neighbourhood U_p serving for *all* functions at once.

A subset $\mathcal{F} \subseteq \mathscr{C}(M)$ is called *pointwise bounded* if for all $p \in M$ one has

$$\sup_{f \in \mathcal{F}} |f(p)| < \infty. \qquad (6.3.5)$$

Note that the value of the supremum still can depend on the point p. In particular, it may well be that there are unbounded functions in a pointwise bounded subset $\mathcal{F}$. With these two notions one can now formulate the Arzelà-Ascoli Theorem as follows:

Theorem 6.3.4 (Arzelà-Ascoli) *Let $(M, \mathcal{M})$ be a compact Hausdorff space and let $\mathcal{F} \subseteq \mathscr{C}(M)$ be a non-empty subset. Then $\mathcal{F}$ is totally bounded iff $\mathcal{F}$ is pointwise bounded and equicontinuous.*

Proof Assume first that $\mathcal{F}$ is pointwise bounded and equicontinuous and let $\epsilon > 0$ be given. Choose $U_p \in \mathfrak{U}(p)$ for every $p \in M$ such that (6.3.3) holds. Then already finitely many $U_{p_1}, \ldots, U_{p_n}$ of these neighbourhoods cover M by compactness. This means

$$|f(p_i) - f(q)| < \epsilon$$

for $q \in U_{p_i}$ and all $f \in \mathcal{F}$. Since every point $p \in M$ is contained in at least one U_{p_i} we get from pointwise boundedness for all $p \in M$ an i such that $p \in U_{p_i}$ and

$$\sup_{f \in \mathcal{F}} |f(p)| \leq \sup_{f \in \mathcal{F}} \left\{ |f(p_i)| + |f(p_i) - f(p)| \right\} \qquad (*)$$

$$\leq \sup_{f \in \mathcal{F}} |f(p_i)| + \epsilon$$

$$\leq \max_{i=1}^{n} \sup_{f \in \mathcal{F}} |f(p_i)| + \epsilon,$$

which is finite and independent of p. This means that $\mathcal{F}$ is bounded with respect to the supremum norm, i.e.

$$\sup_{f \in \mathcal{F}} \|f\|_\infty = \sup_{p \in M} \sup_{f \in \mathcal{F}} |f(p)| \leq R < \infty,$$

where $R = \max_{i=1}^{n} \sup_{f \in \mathcal{F}}(f(p_i)) + \epsilon$ is the bound achieved in $(*)$. Consider now the compact ball $\mathrm{B}_R(0)^{\mathrm{cl}} \subseteq \mathbb{C}$ and the map

$$\Phi : \mathcal{F} \ni f \mapsto (f(p_1), \ldots, f(p_n)) \in \left(\mathrm{B}_R(0)^{\mathrm{cl}} \right)^n.$$

Since $\left(B_R(0)^{cl}\right)^n \subseteq \mathbb{C}^n$ is compact again, the points $\{\Phi(f)\}_{f \in \mathcal{F}}$ have to accumulate in the sense that there must be finitely many $f_1, \ldots, f_N \in \mathcal{F}$ such that for all $f \in \mathcal{F}$ there is a $k \in \{1, \ldots, N\}$ with

$$\|\Phi(f) - \Phi(f_k)\|_{max} < \epsilon, \tag{$**$}$$

where we use the maximum norm of $\mathbb{C}^n$. Indeed, consider the set of points

$$K = \{\Phi(f) \mid f \in \mathcal{F}\}^{cl} \subseteq B_R(0)^{cl},$$

which is compact. Now it is clear that the open balls $\{B_\epsilon(\Phi(f))\}_{f \in \mathcal{F}}$ provide an open cover of K. Hence finitely many $f_1, \ldots, f_N$ suffice for $K \subseteq B_\epsilon(\Phi(f_1)) \cup \cdots \cup B_\epsilon(\Phi(f_N))$. Since for every $f \in \mathcal{F}$ we have $\Phi(f) \in K$ this implies that there is at least one k with $\Phi(f) \in B_\epsilon(\Phi(f_k))$ which is $(**)$. Since we used the maximum norm in $\mathbb{C}^n$ in $(**)$ this is equivalent to the statement that for all f there is a k with

$$|f(p_i) - f_k(p_i)| < \epsilon \tag{$\star$}$$

for all $i = 1, \ldots, n$. Furthermore, for $p \in U_{p_i}$ we have

$$|f(p) - f(p_i)| < \epsilon \quad \text{and} \quad |f_k(p) - f_k(p_i)| < \epsilon \tag{$\star\star$}$$

by equicontinuity. Thus putting $(\star)$ and $(\star\star)$ together gives

$$|f(p) - f_k(p)| \le |f(p) - f(p_i)| + |f(p_i) - f_k(p_i)| + |f_k(p_i) - f_k(p)| < 3\epsilon$$

for all $p \in M$ since every point is in some U_{p_i}. But this means that also $\|f - f_k\|_\infty \le 3\epsilon$ and thus every $f \in \mathcal{F}$ is inside a closed 3ϵ-ball around one of the $f_1, \ldots, f_N$. Since $\epsilon > 0$ was arbitrary, we can rescale appropriately to conclude that $\mathcal{F}$ is totally bounded. For the converse, assume that $\mathcal{F}$ is totally bounded, and let $\epsilon > 0$ be given. Fix $f_1, \ldots, f_N$ with (6.3.1), i.e. for all $f \in \mathcal{F}$ there exists at least one $k = 1, \ldots, N$ with $f \in B_\epsilon(f_k)$ where now $B_\epsilon(\cdot)$ is the metric ball with respect to $\|\cdot\|_\infty$. Hence $\|f - f_k\|_\infty < \epsilon$. For $p \in M$ this means

$$|f(p)| \le |f(p) - f_k(p)| + |f_k(p)| \le \epsilon + \max_{k=1}^{N} |f_k(p)|.$$

Since the right hand side is independent of f, we get

$$\sup_{f \in \mathcal{F}} |f(p)| \le \epsilon + \max_{k=1}^{N} |f_k(p)| < \infty,$$

and thus $\mathcal{F}$ is pointwise bounded. Now let $p \in M$ and choose $U_{p,k} \in \mathfrak{U}(p)$ for $k = 1, \ldots, N$ such that for all $q \in U_{p,k}$ we have

$$|f_k(p) - f_k(q)| < \epsilon, \qquad\qquad (\text{☺})$$

which is possible since f_k is continuous at p. Taking $U_p = U_{p,1} \cap \cdots \cap U_{p,N} \in \mathfrak{U}(p)$ we get the continuity estimate (☺) for all $q \in U_p$ and all $k = 1, \ldots, N$ at once. Now we have for $f \in \mathcal{F}$ and $q \in U_p$

$$|f(p) - f(q)| \le |f(p) - f_k(p)| + |f_k(p) - f_k(q)| + |f_k(q) - f(q)| < 3\epsilon$$

by (☺) and k chosen such that we can apply $\|f - f_k\|_\infty < \epsilon$ by total boundedness. But this is the equicontinuity we want to show as $\epsilon > 0$ is arbitrary. $\square$

The Arzelà-Ascoli Theorem has several corollaries which might provide a more familiar formulation: since $\mathscr{C}(M)$ is a complete metric space, a subset $\mathcal{F} \subseteq \mathscr{C}(M)$ is compact iff it is closed and totally bounded by Exercise 5.5.7 as well as Proposition 6.3.1. Since $\mathcal{F}$ is totally bounded iff $\mathcal{F}^{\mathrm{cl}}$ is totally bounded we can always pass to a closed subset without further difficulties. This gives the following first corollary of the Arzelà-Ascoli Theorem:

Corollary 6.3.5 *Let $(M, \mathcal{M})$ be a compact Hausdorff space. Then $\mathcal{F} \subseteq \mathscr{C}(M)$ is pointwise bounded and equicontinuous iff $\mathcal{F}^{\mathrm{cl}}$ is compact.*

This way we get a rather explicit and simple characterization of the compact subsets of $\mathscr{C}(M)$. Since compactness is equivalent to sequential compactness for a metric space by Proposition 5.4.12 we get the following statement:

Corollary 6.3.6 *Let $(M, \mathcal{M})$ be a compact Hausdorff space and let $\mathcal{F} \subseteq \mathscr{C}(M)$ be a subset. Then $\mathcal{F}$ is pointwise bounded and equicontinuous iff every sequence in $\mathcal{F}$ has a convergent subsequence.*

Of course, the limit of this subsequence is in $\mathcal{F}^{\mathrm{cl}}$ but not necessarily in $\mathcal{F}$.

Remark 6.3.7 There are further generalizations of the Arzelà -Ascoli Theorem: one can relax the compactness assumption on $(M, \mathcal{M})$ to local compactness and one can replace the target $\mathbb{C}$ by some much more general target: namely by a *uniform space* $\mathcal{X}$, a notion which we have not discussed. Putting the correct topology on $\mathscr{C}(M, \mathcal{X})$ still allows for a characterization of the compact subsets of $\mathscr{C}(M, \mathcal{X})$, see e.g. the discussion in [27, Sect. 14C] or [17, Chap. 7, Theorem 18].

The Arzelà -Ascoli Theorem is one of the very fundamental results needed for various applications in functional analysis, differential equations, and holomorphic function theory. We mention just one application: Montel's Theorem. We assume some familiarity with holomorphic functions for the rest of this section.

Consider a non-empty open subset $X \subseteq \mathbb{C}$ and denote the holomorphic functions on X by $\mathcal{O}(X)$. Since every holomorphic function is continuous we can view $\mathcal{O}(X)$ as a subspace of $\mathscr{C}(X)$. This way, $\mathcal{O}(X)$ inherits the topology of locally uniform convergence, see Exercise 6.4.3. It is a well-known result from holomorphic function theory that the locally uniform limit of holomorphic functions is again holomorphic.

With other words, $\mathscr{O}(X) \subseteq \mathscr{C}(X)$ is a *closed* subspace, see also Exercise 6.4.4, (iv). Since $X \subseteq \mathbb{C}$ is non-compact, $\mathscr{C}(X)$ and hence $\mathscr{O}(X)$ are no longer Banach spaces. Nevertheless, the topology of $\mathscr{C}(X)$ is still first countable by Proposition 6.2.10 and, in fact, *metrizable*: there exists a metric d on $\mathscr{C}(X)$ such that $(\mathscr{C}(X), d)$ is a complete metric space with the topology being the one of locally uniform convergence. Hence also $\mathscr{O}(X)$ becomes a complete metric space, see also Exercise 6.4.9.

The following lemma gives a complete and very simple characterization of equicontinuous sets of *holomorphic* functions:

Lemma 6.3.8 *Let $X \subseteq \mathbb{C}$ be open and let $\mathcal{F} \subseteq \mathscr{O}(X)$ be a locally bounded set of holomorphic functions. Then $\mathcal{F}$ is equicontinuous.*

Proof Let $\epsilon > 0$ and $z \in X$ be given. Then we find a radius $R > 0$ such that $B_R(z)^{\mathrm{cl}} \subseteq X$ and $\mathcal{F}$ is bounded on this compact subset, i.e. there is a $C_R > 0$ such that $|f(w)| \leq C_R$ for all $w \in B_R(z)^{\mathrm{cl}}$. Let $w \in B_{\frac{R}{2}}(z)$. Then we can estimate the difference of $f(w)$ and $f(z)$ by first integrating the derivative f' along some path joining w and z. This gives

$$|f(w) - f(z)| = \left| \int_z^w f'(u) \mathrm{d}u \right| \leq |z - w| \sup_{u \in B_{\frac{R}{2}}(z)} |f'(u)|. \qquad (*)$$

In a second step we estimate the derivative by means of the Cauchy formula

$$|f'(u)| = \left| \frac{1}{2\pi \mathrm{i}} \int_{\partial B_R(z)} \frac{f(\zeta) \mathrm{d}\zeta}{(u - \zeta)^2} \right| \leq \frac{1}{2\pi} 2\pi R \frac{C_R}{(\frac{R}{2})^2} = \frac{4C_R}{R}, \qquad (**)$$

where the contour integral is performed along the boundary line of the disc $B_R(z) \subseteq X$, having length $2\pi R$. Note that $|u - \zeta|$ is at least $\frac{R}{2}$ since $u \in B_{\frac{R}{2}}(z)$ while $\zeta \in \partial B_R(z)$, justifying this estimate. Without restriction, we can assume $C_R > \epsilon$. Then $\delta = \frac{R\epsilon}{4C_R} > 0$ is clearly small enough such that $B_\delta(z) \subseteq B_{\frac{R}{2}}(z)$, since $\delta < \frac{R}{4}$. Now for $w \in B_\delta(z)$ we have

$$|f(w) - f(z)| \overset{(*)}{\leq} |z - w| \sup_{u \in B_{\frac{R}{2}}(z)} |f'(z)| \overset{(**)}{\leq} \delta \frac{4C_R}{R} = \epsilon,$$

which is independent of f. This shows that the neighbourhood $B_\delta(z)$ of z serves as the neighbourhood needed to conclude the equicontinuity of $\mathcal{F}$ at z. Since $z \in X$ was arbitrary, we have equicontinuity of $\mathcal{F}$ on X. $\qquad \square$

Using this characterization of equicontinuity for holomorphic functions the Arzelà-Ascoli Theorem provides now a simple way of describing compact subsets of $\mathscr{O}(X)$:

Theorem 6.3.9 (Montel) *Let $X \subseteq \mathbb{C}$ be a non-empty open subset and endow the holomorphic functions $\mathscr{O}(X)$ with the topology of locally uniform convergence. Then a subset $\mathcal{F} \subseteq \mathscr{O}(X)$ is compact iff it is closed and locally bounded.*

Proof Suppose first that $\mathcal{F}$ is compact. Since the topology of $\mathcal{O}(X)$ is Hausdorff, $\mathcal{F}$ is closed. Since the seminorms $\| \cdot \|_K : \mathcal{O}(X) \longrightarrow [0, \infty)$ are continuous for all compact subsets $K \subseteq X$, they map the compact subset to a compact and hence bounded subset of $[0, \infty)$ by Proposition 5.2.6. But this is precisely the local boundedness of $\mathcal{F}$. For the (non-trivial) reverse direction we assume that $\mathcal{F}$ is closed and locally bounded. Then the previous lemma shows that $\mathcal{F}$ is equicontinuous everywhere. The closedness of $\mathcal{F}$ immediately implies that

$$\mathcal{F}_K = \left\{ f\big|_K \in \mathscr{C}(K) \mid f \in \mathcal{F} \right\}$$

is closed in $\mathscr{C}(K)$, e.g. using an argument via convergence of sequences which suffices as we are now in a Banach space situation anyway. Thus the Arzelà-Ascoli Theorem in form of Corollary 6.3.6 shows that every sequence in $\mathcal{F}_K$ has a convergent subsequence with limit still in $\mathcal{F}_K$. The following standard argument shows that a sequence in $\mathcal{F}$ has a convergent subsequence. Let $(f_n)_{n\in\mathbb{N}}$ be a sequence of holomorphic functions in $\mathcal{F}$ and choose an exhausting sequence of compact subsets $\ldots \subseteq K_m \subseteq K_{m+1}^\circ \subseteq K_{n+1} \subseteq \ldots \subseteq X$ for X according to Proposition 6.2.10. Since K_1 is compact, the restricted sequence $\left(f_n\big|_{K_1} \right)_{n\in\mathbb{N}}$ has a convergent subsequence according to the compactness of $\mathcal{F}_{K_1}$. Denote this subsequence by $\left(f_n^{(1)} \right)_{n\in\mathbb{N}} = (f_{i_n})_{n\in\mathbb{N}}$. Now we consider $\left(f_n^{(1)}\big|_{K_2} \right)_{n\in\mathbb{N}}$ for which we find a convergent subsequence with respect to $\| \cdot \|_{K_2}$, denoted by $\left(f_n^{(2)} \right)_{n\in\mathbb{N}}$ and so on. This gives us a sequence of subsequences $\left(f_n^{(m)} \right)_{n\in\mathbb{N}}$ of the original sequence such that $\left(f_n^{(m)} \right)_{n\in\mathbb{N}}$ converges with respect to $\| \cdot \|_{K_m}$ and $\left(f_n^{(m+1)} \right)_{n\in\mathbb{N}}$ is a subsequence of $\left(f_n^{(m)} \right)$. Define now $g_n = f_n^{(n)}$ which is the diagonal subsequence. We claim that $(g_n)_{n\in\mathbb{N}}$ converges with respect to all seminorms $\| \cdot \|_{K_m}$. Indeed, for $n \geq m$ the sequence $(g_n)_{n\in\mathbb{N}}$ is a subsequence of the sequence $\left(f_n^{(m)} \right)_{n\in\mathbb{N}}$ which converges with respect to $\| \cdot \|_{K_m}$. Hence also $(g_n)_{n\in\mathbb{N}}$ converges with respect to $\| \cdot \|_{K_m}$. The closedness of $\mathcal{F}$ shows that the limit g of this sequence still belongs to $\mathcal{F}$. Thus $\mathcal{F}$ is sequentially compact which is the same thing as compact since the topology of $\mathcal{F}$ is metric, see again Exercise 6.4.9. $\qquad\qquad\square$

Corollary 6.3.10 *Let $\mathcal{F} \subseteq \mathcal{O}(X)$ be a family of holomorphic functions which is locally bounded. Then every sequence in $\mathcal{F}$ has a convergent subsequence with respect to locally uniform convergence. The limit will be a holomorphic function on X in the closure $\mathcal{F}^{\mathrm{cl}}$ of $\mathcal{F}$.*

Remark 6.3.11 From the point of view of a locally convex space a locally bounded subset $\mathcal{F} \subseteq \mathcal{O}(X)$ in the sense above is just a *bounded* subset with respect to all the seminorms $\| \cdot \|_K$. Hence the locally convex space $(\mathcal{O}(X), d)$ has the *Heine-Borel property*: the closed and bounded subsets are precisely the compact subsets. This is in so far very remarkable as for a Banach space this property holds iff the underlying vector space is *finite-dimensional*. In this sense the infinite-dimensional space $\mathcal{O}(X)$ behaves very much like a finite-dimensional one.

6.4 Exercises

Exercise 6.4.1 (Approximation of the square root) Show that there is a sequence of real polynomials $p_n \in \mathbb{R}[x]$ with $p_n(0) = 0$ which approximate the square root function $x \mapsto \sqrt{x}$ uniformly on the unit interval $[0, 1]$. Extend this result to an arbitrary interval $[0, a]$ with $a > 0$.

Hint: Define the polynomials recursively by

$$p_0(x) = 0 \quad \text{and} \quad p_{n+1}(x) = p_n(x) + \frac{1}{2}\left(x - p_n^2(x)\right), \tag{6.4.1}$$

and show that beside $p_n(0) = 0$ one has the estimates

$$p_n(x) \geq 0 \quad \text{and} \quad 0 \leq \sqrt{x} - p_n(x) \leq \frac{2\sqrt{x}}{2 + n\sqrt{x}} \tag{6.4.2}$$

for all $x \in [0, 1]$. Use this to finish the proof.

Exercise 6.4.2 (Normed *-algebra) Let $\mathcal{A}$ be a normed *-algebra over $\mathbb{C}$. Show that all structure maps

$$\mathcal{A} \times \mathcal{A} \ni (a, b) \mapsto a + b \in \mathcal{A},$$
$$\mathbb{C} \times \mathcal{A} \ni (z, a) \mapsto za \in \mathcal{A},$$
$$\mathcal{A} \times \mathcal{A} \ni (a, b) \mapsto ab \in \mathcal{A},$$
$$\mathcal{A} \ni a \mapsto a^* \in \mathcal{A}$$

are continuous.

Exercise 6.4.3 (Locally bounded functions) Let M be a locally compact space and let $A \subseteq M$ be a subset. Then one defines

$$\|f\|_A = \sup_{p \in A} |f(p)| \tag{6.4.3}$$

for a function $f : M \longrightarrow \mathbb{C}$. A function f is called *locally bounded*, if for every point $p \in M$ there is a neighbourhood $U \in \mathfrak{U}(p)$ such that $\|f\|_U < \infty$. The set of locally bounded functions will be denoted by $\mathcal{B}_{\mathrm{loc}}(M)$.

  (i) Show that $\mathcal{B}_{\mathrm{loc}}(M)$ is a vector space (and even a *-algebra) with respect to the pointwise operations.
 (ii) Show that $f \in \mathcal{B}_{\mathrm{loc}}(M)$ iff $\|f\|_K < \infty$ for all compact subsets $K \subseteq M$. Show also that $\| \cdot \|_K$ is a well-defined seminorm on $\mathcal{B}_{\mathrm{loc}}(M)$ for every compact subset $K \subseteq M$.
(iii) Let $(f_i)_{i \in I}$ be a net of locally bounded functions which converges locally uniformly to a function f, i.e. for every point $p \in M$ there is a neighbourhood $U \in \mathfrak{U}(p)$ such that on this neighbourhood $f_i \longrightarrow f$ uniformly. Show that this is equivalent to the statement that $\|f_i - f\|_K \longrightarrow 0$ for every compact subset $K \subseteq M$. Conclude that in this case $f \in \mathcal{B}_{\mathrm{loc}}(M)$.

(iv) Let $(f_i)_{i \in I}$ be a net in $\mathscr{B}_{\text{loc}}(M)$ such that for every $p \in M$ there is a neighbourhood $U \in \mathfrak{U}(p)$ with the property that for every $\epsilon > 0$ one finds an index $i \in I$ with

$$\| f_j - f_k \|_U < \epsilon \qquad (6.4.4)$$

for all $j, k \in I$ with $j, k \succcurlyeq i$. Such a net is called a *Cauchy net*. Show that a net $(f_i)_{i \in I}$ in $\mathscr{B}_{\text{loc}}(M)$ is a Cauchy net iff the above Cauchy property holds for all seminorms $\| \cdot \|_K$. Show that a Cauchy net $(f_i)_{i \in I}$ in $\mathscr{B}_{\text{loc}}(M)$ converges to a function $f \in \mathscr{B}_{\text{loc}}(M)$ in the sense of part (iii). Thus the vector space $\mathscr{B}_{\text{loc}}(M)$ is *complete* with respect to locally uniform convergence.

(v) Show that $\mathscr{C}(M) \subseteq \mathscr{B}_{\text{loc}}(M)$ is a subspace (even a *-subalgebra).

(vi) Show that $\mathscr{C}(M) \subseteq \mathscr{B}_{\text{loc}}(M)$ is closed with respect to locally uniform convergence.

(vii) Conclude that $\mathscr{C}(M)$ is complete with respect to locally uniform convergence.

Let us conclude this exercise with some remarks. As for a normed space we can define a topology out of the seminorms $\{ \| \cdot \|_K \}_{K \subseteq M \text{ compact}}$ for $\mathscr{B}_{\text{loc}}(M)$: one defines the open balls

$$\mathrm{B}_{\| \cdot \|_K, r}(f) = \left\{ g \in \mathscr{B}_{\text{loc}}(M) \mid \| f - g \|_K < r \right\} \qquad (6.4.5)$$

to be open and considers the generated topology. With some slightly larger effort one can show that the above notions of locally uniform convergence are indeed the notions of convergence in the topological space $\mathscr{B}_{\text{loc}}(M)$. The main new point is that it will depend on the nature of M whether or not one will need uncountably many seminorms $\| \cdot \|_K$ in order to obtain a neighbourhood basis of a point: the resulting topology may fail to be first countable. This happens whenever there are "too many" compact subsets of M. Thus the usage of nets instead of sequences will be inevitable. Finally, let us note that this construction of a topology out of seminorms is of course very generic and can be done for other vector spaces different from $\mathscr{B}_{\text{loc}}(M)$ as well. The resulting theory of *locally convex spaces* is a very rich and interesting branch in functional analysis , see e.g. [14,18,19,34] for a further reading.

Exercise 6.4.4 (Variations on the Theorem of Stone-Weierstraß)

(i) Show that the polynomials are dense in the continuous functions on a compact subset $K \subseteq \mathbb{R}^n$.

(ii) Let $(M, \mathcal{M})$ be a compact Hausdorff space and let $\mathcal{A} \subseteq \mathscr{C}(M)$ be a point-separating *-subalgebra with the property that for every $p \in M$ there is a function $g_p \in \mathcal{A}$ with $g_p(p) \neq 0$. Show that $\mathcal{A}$ is dense in $\mathscr{C}(M)$.
Hint: Let $f \in \mathscr{C}(M)$ be given. Show that there is a function $\varepsilon \in \mathcal{A}$ with $\varepsilon > 0$. Choose now for two points $p \neq q$ a function $g = \overline{g} \in \mathcal{A}$ with $g(p) \neq g(q)$ and consider

$$h = \frac{f(p) - f(q)}{g(p) - g(q)} g - \frac{f(p)g(q) - f(q)g(p)}{g(p) - g(q)} \frac{\varepsilon(q) + \varepsilon(p) - \varepsilon}{\varepsilon(p)\varepsilon(q)} \varepsilon.$$

Show that $h \in \mathcal{A}$ and $h(p) = f(p)$ and $h(q) = f(q)$. Finally, show that for $p = q$ there is a function $h \in \mathcal{A}$ with $h(p) = f(p)$. Why is this all one has to show?

(iii) Consider a unital point-separating $\mathbb{R}$-subalgebra $\mathcal{A} \subseteq \mathscr{C}(M, \mathbb{R})$ for a locally compact Hausdorff space $(M, \mathcal{M})$. Show that $\mathcal{A}^{\mathrm{cl}} = \mathscr{C}(M, \mathbb{R})$ with respect to locally uniform convergence.

Hint: Consider the $\mathbb{C}$-span $\mathcal{A}_{\mathbb{C}}$ of $\mathcal{A}$ inside $\mathscr{C}(M)$.

(iv) Consider the open unit disc $\mathbb{D} \subseteq \mathbb{C}$ and view the holomorphic polynomials $\mathcal{A} = \mathbb{C}[z]$ as continuous functions on $\mathbb{D}$. Show that $\mathcal{A}$ is a unital point-separation subalgebra which is *not* dense in $\mathscr{C}(\mathbb{D})$ with respect to locally uniform convergence. What is the closure $\mathcal{A}^{\mathrm{cl}}$ of this subalgebra inside $\mathscr{C}(\mathbb{D})$?

(v) Show that the existence of a unit can not be abandoned easily: the polynomials $x\mathbb{C}[x]$ with vanishing constant term satisfy all requirements of the Stone-Weierstraß Theorem except of having a unit. Show that the closure of $x\mathbb{C}[x]$ in $\mathscr{C}([0, 1])$ is not all of $\mathscr{C}([0, 1])$.

Exercise 6.4.5 (Sum of locally finitely supported functions) Let $(M, \mathcal{M})$ be a locally compact Hausdorff space and let $\{f_i\}_{i \in I}$ be a collection of continuous functions with locally finite supports. Discuss in which sense the series

$$f = \sum_{i \in I} f_i \tag{6.4.6}$$

converges with respect to the topology of locally uniform convergence in $\mathscr{C}(M)$.

Exercise 6.4.6 (Not locally finite open cover) Give an example of an open cover of $\mathbb{R}^n$ which is not locally finite.

Exercise 6.4.7 (Devil's staircase) Consider again the Cantor set $C \subseteq [0, 1]$ as in Exercise 2.7.24 with the corresponding closed sets $A_n \subseteq [0, 1]$ such that $C = \bigcap_{n=1}^{\infty} A_n$. Now define continuous functions $f_n : [0, 1] \longrightarrow [0, 1]$ as follows. The first function f_1 is constant on $[0, 1] \setminus A_1$ and equal to $\frac{1}{2}$ there. For $x \in [0, \frac{1}{3}]$ we set $f_1(x) = \frac{3}{2}x$ and for $x \in [\frac{2}{3}, 1]$ we set $f_1(x) = \frac{3}{2}x - \frac{1}{2}$. In the next iteration we keep the values where f_n is already constant and add constant pieces for every new points in $[0, 1] \setminus A_n$ in the middle between the previous constant pieces. On A_n itself we connect the constant plateaus by linear functions in a continuous way, see Fig. 6.4 for f_3.

(i) Show that the sequence f_n converges uniformly to a continuous function $f_\infty : [0, 1] \longrightarrow [0, 1]$. This limit function is called the *devil's staircase*.

(ii) Show that f_∞ is differentiable on $[0, 1] \setminus C$ with zero derivative.

In view of the characterization of the Cantor set C in Exercise 2.7.24, (iii) and (iv), the devil's staircase is differentiable on a rather large subset of the unit interval. Even though it has zero derivative there, it climbs from 0 to 1, quite challenging to visualize. Nevertheless, this function can be uniformly approximated by polynomials according to Exercise 6.4.4, (i).

Exercise 6.4.8 (Distance and exhausting compact subsets) Consider a metric space (M, d) and let $A \subseteq M$ be a non-empty closed subset.

Fig. 6.4 The function f_3

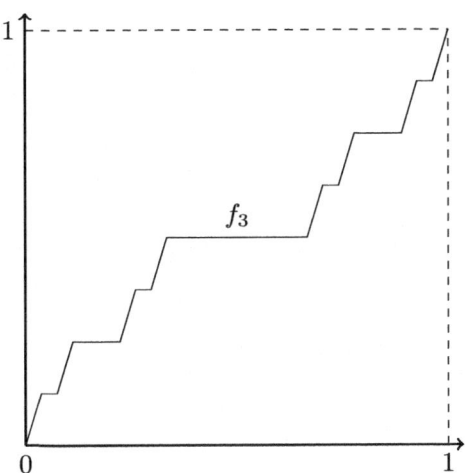

(i) For $p \in M$ one defines the *distance* of p to A by

$$\mathrm{dist}(p, A) = \inf\{d(p, q) \mid q \in A\}. \tag{6.4.7}$$

Show that $p \in A$ iff $\mathrm{dist}(p, A) = 0$. Show also that the function $p \mapsto \mathrm{dist}(p, A)$ is continuous. Argue that it is reasonable to set $\mathrm{dist}(p, \emptyset) = +\infty$.
Hint: Show first that $\mathrm{dist}(p, A) \leq \mathrm{dist}(p', A) + d(p, p')$ for $p, p' \in M$.

(ii) Consider now a non-empty open subset $\mathcal{O} \subseteq M$ and set $A = M \setminus \mathcal{O}$. Consider

$$K_n = \left\{ p \in M \mid d(p, p_0) \leq n \text{ and } \mathrm{dist}(p, A) \geq \tfrac{1}{n} \right\}, \tag{6.4.8}$$

where $p_0 \in M$ is a chosen point. Show that $K_n \subseteq K_{n+1}^\circ \subseteq K_{n+1} = K_{n+1}^{\mathrm{cl}}$ for all $n \in \mathbb{N}$. Show also that $\bigcup_{n \in \mathbb{N}} K_n = \mathcal{O}$. This way one has found an exhausting sequence of closed subsets of $\mathcal{O}$.

(iii) Now assume in addition that all closed balls around p_0 are compact. Show that in this case K_n is compact for all $n \in \mathbb{N}$. This provides a more intuitive construction of an exhausting sequence of compact subsets than Proposition 6.2.10, provided the extra assumption is fulfilled.

(iv) Construct a second countable metric space (M, d) such that around every point one has a non-compact closed ball.

Exercise 6.4.9 ($\mathscr{C}(M)$ is metrizable) Consider a second countable, locally compact Hausdorff space $(M, \mathcal{M})$. Choose an exhausting sequence $(K_n)_{n \in \mathbb{N}}$ of compact subsets according to Proposition 6.2.10, (ii). Consider then the supremum seminorms

$$\|f\|_n = \sup_{p \in K_n} |f(p)|, \tag{6.4.9}$$

and define

$$d(f, g) = \sum_{n=1}^{\infty} \frac{1}{2^n} \frac{\|f - g\|_n}{1 + \|f - g\|_n} \tag{6.4.10}$$

for $f, g \in \mathscr{C}(M)$.

(i) Show that the seminorms $\| \cdot \|_n$ already determine the topology of locally uniform convergence for $\mathscr{C}(M)$ in the sense that the open balls with respect to these seminorms form a basis of the topology.

(ii) Show that d yields a complete metric for $\mathscr{C}(M)$ which induces again the same topology of locally uniform convergence.

Exercise 6.4.10 (Local units) Let $(M, \mathcal{M})$ be a locally compact Hausdorff space. Show that there is a family $\{\chi_i\}_{i \in I}$ of real-valued continuous functions $\chi_i = \overline{\chi_i} \in \mathscr{C}_0(M)$ with compact support having the following property: for $f_1, \ldots, f_n \in \mathscr{C}_0(M)$ there is an index $i \in I$ with

$$f_\alpha \chi_i = f_\alpha \tag{6.4.11}$$

for all $\alpha = 1, \ldots, n$. Such a family $\{\chi_i\}_{i \in I}$ is also called a *local unit* for the (in general non-unital) algebra $\mathscr{C}_0(M)$. Show that I can be chosen to be countable if $(M, \mathcal{M})$ is second countable.
Hint: Use Corollary 6.1.3 and Proposition 6.2.10.

Exercise 6.4.11 (Existence of a proper function) Let $(M, \mathcal{M})$ be a second countable locally compact Hausdorff space. Show that there is a proper function $f : M \longrightarrow \mathbb{R}$.
Hint: Take an exhausting sequence $\{K_n\}_{n \in \mathbb{N}}$ of compact subsets of M as in Proposition 6.2.10, (ii). Consider functions $f_n \in \mathscr{C}(M)$ with supp $f_n \subseteq K_{n+1}^\circ \setminus K_{n-2}$ such that $f_n|_{K_n \setminus K_{n-1}^\circ} = n$ and $0 \leq f_n(p) \leq n$ everywhere. Why do such functions exist? Consider then $f = \sum_{n \in \mathbb{N}} f_n$.

Chapter 7
Baire's Theorem

In a topological space we have an intuitive understanding of "small" and "large" subsets. One would consider a dense subset "large" compared to a non-dense subset which in turn can still be considered to be "large" compared to a nowhere dense subset. The notion of meager sets gives a more refined notion for this problem and Baire's Theorem helps to characterize meager subsets in particular situations. Again, as for the Stone-Weierstraß Theorem or the Arzelà-Ascoli Theorem there are important applications of Baire's Theorem to various areas in mathematics, in particular in functional analysis.

7.1 Meager Subsets and Baire Spaces

Recall that a subset A of a topological space $(M, \mathcal{M})$ is called nowhere dense if the open interior of the closure of A is empty, i.e.

$$(A^{\mathrm{cl}})^\circ = \emptyset. \tag{7.1.1}$$

A slight generalization of this is the notion of a meager set:

Definition 7.1.1 (*Meager Set*) Let $(M, \mathcal{M})$ be a topological space. Then $A \subseteq M$ is called meager (or of first category) if A is a countable union of nowhere dense subsets. A non-meager subset is also called of second category or residual.

Example 7.1.2 (Meager sets)

  (i) $\mathbb{Q}$ with its standard topology is meager in $\mathbb{R}$ as $\mathbb{Q}$ is the countable union of the nowhere dense subsets $\{r\}_{r \in \mathbb{Q}}$. For the same reason, $\mathbb{Q}$ is meager in $\mathbb{Q}$. Note that $\mathbb{Q}$ is dense in $\mathbb{R}$, so a meager subset can very well be dense.
  (ii) $\mathbb{R}$ is not meager in $\mathbb{R}$. In fact, this is not completely obvious: it will follow easily from Baire's Theorem later but a direct proof requires some more work, see e.g. the discussion in [27, 13.A19]. Nevertheless, $\mathbb{R}$ viewed as subset of $\mathbb{C}$

© Springer International Publishing Switzerland 2014
S. Waldmann, *Topology*, DOI 10.1007/978-3-319-09680-3_7

is meager, in fact even nowhere dense. Thus it is very important to specify the ambient space.

(iii) The Cantor set $C \subseteq [0, 1]$ is nowhere dense in $[0, 1]$ and hence meager. In particular, a meager subset needs not to be countable.

(iv) Let $(M, \mathcal{M})$ be a T_1-space such that a single point is not open, i.e. $\{p\}^\circ = \emptyset$ for all $p \in M$. Since $\{p\}^{\mathrm{cl}} = \{p\}$ we conclude that in such a space any countable subset is meager, generalizing the first example.

The following proposition collects some first properties of meager subsets:

Proposition 7.1.3 *Let $(M, \mathcal{M})$ be a topological space.*

(i) A subset of a nowhere dense subset is again nowhere dense.
(ii) A finite union of nowhere dense subsets is again nowhere dense.
(iii) A subset of a meager subset is again meager.
(iv) A countable union of meager subsets is again meager.

Proof Suppose that A is nowhere dense and $B \subseteq A$. Then $B^{\mathrm{cl}} \subseteq A^{\mathrm{cl}}$ and hence $(B^{\mathrm{cl}})^\circ \subseteq (A^{\mathrm{cl}})^\circ = \emptyset$ shows that B is nowhere dense, too. From Exercise 2.7.12 we know that A is nowhere dense iff $M \setminus A^{\mathrm{cl}}$ is dense (and of course open). Now let $A_1, \ldots, A_n$ be nowhere dense and consider $A = A_1 \cup \cdots \cup A_n$. Then $M \setminus A^{\mathrm{cl}} = M \setminus (A_1 \cup \cdots \cup A_n)^{\mathrm{cl}} = M \setminus (A_1^{\mathrm{cl}} \cup \cdots \cup A_n^{\mathrm{cl}}) = M \setminus A_1^{\mathrm{cl}} \cap \cdots \cap M \setminus A_n^{\mathrm{cl}}$. Let $p \in M$ and $U \in \mathfrak{U}(p)$ be an open neighbourhood of p. Since $M \setminus A_1^{\mathrm{cl}}$ is dense there exists a point $q_1 \in U \cap (M \setminus A_1^{\mathrm{cl}}) \neq \emptyset$. Since $M \setminus A_1^{\mathrm{cl}}$ is open, the subset $U \cap (M \setminus A_1^{\mathrm{cl}})$ is an open neighbourhood of q_1. Since $M \setminus A_2^{\mathrm{cl}}$ is dense we have $U \cap (M \setminus A_1^{\mathrm{cl}}) \cap (M \setminus A_2^{\mathrm{cl}}) \neq \emptyset$. We choose $q_2 \in U \cap (M \setminus A_1^{\mathrm{cl}}) \cap (M \setminus A_2^{\mathrm{cl}})$ and proceed by induction to obtain a $q_n \in U \cap (M \setminus A_1^{\mathrm{cl}}) \cap \cdots \cap (M \setminus A_n^{\mathrm{cl}})$ showing that this is non-empty. This shows that $U \cap (M \setminus A^{\mathrm{cl}})$ is non-empty for all $U \in \mathfrak{U}(p)$ and all $p \in M$. Hence $M \setminus A^{\mathrm{cl}}$ is dense which means that A is nowhere dense. Now let $A = \bigcup_{n=1}^\infty A_n$ be a meager subset where A_n is nowhere dense. If $B \subseteq A$ then $B = \bigcup_{n=1}^\infty (A_n \cap B)$ and $A_n \cap B \subseteq A_n$ is nowhere dense by the first part. Thus B is meager again. Finally let $A_n = \bigcup_{m=1}^\infty A_{nm}$ be a sequence of meager subsets with A_{nm} being nowhere dense. Then

$$\bigcup_{n=1}^\infty A_n = \bigcup_{n,m=1}^\infty A_{nm}$$

is still a countable union of nowhere dense subsets and thus meager, too. □

Note that a countable union of nowhere dense subsets may very well be even dense, see Example 7.1.2, (i). By Exercise 2.7.12 we see that if $\mathcal{O} \subseteq M$ is open and dense then $A = M \setminus \mathcal{O}$ is closed and nowhere dense. Thus the second part has the following corollary:

Corollary 7.1.4 *Let $(M, \mathcal{M})$ be a topological space and let $\mathcal{O}_1, \ldots, \mathcal{O}_n \subseteq M$ be open and dense subsets. Then also $\mathcal{O}_1 \cap \cdots \cap \mathcal{O}_n$ is open and dense.*

The following proposition characterizes a situation where we can extend the above statement to countably many open dense subsets: their intersection will in general be no longer open but one can say something whether or not it is still dense:

Proposition 7.1.5 *Let $(M, \mathcal{M})$ be a topological space. Then the following statements are equivalent:*

(i)   Any countable union of closed subsets of M without inner points has no inner points.

(ii) Any countable intersection of open dense subsets of M is dense.

(iii) Every non-empty open subset of M is not meager.

(iv) The complement of every meager subset of M is dense.

Proof We prove (i) $\Rightarrow$ (ii) $\Rightarrow$ (iii) $\Rightarrow$ (iv) $\Rightarrow$ (i). Assume (i) and let $\{\mathcal{O}_n\}_{n\in\mathbb{N}}$ be a countable family of open dense subsets. Clearly, we can assume it to be infinite in view of Corollary 7.1.4. Then $A_n = M \setminus \mathcal{O}_n$ is closed and nowhere dense, i.e. $A_n^\circ = \emptyset$ since $A_n = A_n^{\mathrm{cl}}$. By (i) and Proposition 2.3.11, (v), we get

$$\left(\bigcap_{n\in\mathbb{N}} \mathcal{O}_n\right)^{\mathrm{cl}} = \left(\bigcap_{n\in\mathbb{N}}(M \setminus A_n)\right)^{\mathrm{cl}}$$
$$= \left(M \setminus \bigcup_{n\in\mathbb{N}} A_n\right)^{\mathrm{cl}}$$
$$= M \setminus \left(\bigcup_{n\in\mathbb{N}} A_n\right)^\circ$$
$$= M,$$

and thus (ii) follows. Next, let $\mathcal{O} \subseteq M$ be open and non-empty. We assume that $\mathcal{O}$ is meager, i.e. of the form $\mathcal{O} = \bigcup_{n\in\mathbb{N}} A_n$ with nowhere dense subsets A_n. Hence $M \setminus A_n^{\mathrm{cl}}$ is dense and open. Thus $\bigcap_{n\in\mathbb{N}}(M \setminus A_n^{\mathrm{cl}})$ is still dense by (ii). This gives

$$\emptyset = M \setminus \left(\bigcap_{n\in\mathbb{N}}(M \setminus A_n^{\mathrm{cl}})\right)^{\mathrm{cl}}$$
$$= M \setminus \left(M \setminus \bigcup_{n\in\mathbb{N}} A_n^{\mathrm{cl}}\right)^{\mathrm{cl}}$$
$$= M \setminus \left(M \setminus \left(\bigcup_{n\in\mathbb{N}} A_n^{\mathrm{cl}}\right)^\circ\right)$$
$$= \left(\bigcup_{n\in\mathbb{N}} A_n^{\mathrm{cl}}\right)^\circ. \tag{$*$}$$

But $\mathcal{O} = \bigcup_{n\in\mathbb{N}} A_n \subseteq \bigcup_{n\in\mathbb{N}} A_n^{\mathrm{cl}}$ is open and thus the open interior of the right hand side of $(*)$ contains at least $\mathcal{O}$, a contradiction. Now assume (iii) and let $A \subseteq M$ be a meager subset such that $M \setminus A$ is not dense. This means that $(M \setminus A)^{\mathrm{cl}} = M \setminus A^\circ$ is not M and thus $A^\circ \neq \emptyset$. Since $A^\circ \subseteq A$ is a subset of a meager set, A° is meager itself and hence we get a contradiction to our assumption (iii). Finally, assume (iv) and let $A = \bigcup_{n\in\mathbb{N}} A_n$ with $A_n^{\mathrm{cl}} = A_n$ and $A_n^\circ = \emptyset$. By definition, A is meager. By (iv) we have $M = (M \setminus A)^{\mathrm{cl}} = M \setminus A^\circ$ showing $A^\circ = \emptyset$ as wanted. $\square$

Clearly, any of these equivalent properties of $(M, \mathcal{M})$ is very desirable as it provides a natural generalization of the statement of Corollary 7.1.4. More informally, they characterize meager subsets as being rather "small" and complements of meager sets as being rather "fat". Topological spaces having this property are called Baire spaces:

Definition 7.1.6 (*Baire space*) Let $(M, \mathcal{M})$ be a topological space. Then M is called a Baire space if one (and hence all) of the statements in Proposition 7.1.5 are fulfilled.

Example 7.1.7 (Baire spaces I)

 (i) The indiscrete topology on a set M will make it a Baire space. Indeed, the only open dense subset is M itself since the other open set $\emptyset$ is not dense. Thus by Proposition 7.1.5, (ii), the indiscrete space $(M, \mathcal{M}_{\text{indiscrete}})$ is a Baire space.
 (ii) The discrete topology on a set M will also turn it into a Baire space. Since every closed subset is also open, the only closed subset without inner points is $\emptyset$. Thus the first characterization in Proposition 7.1.5 shows that $(M, \mathcal{M}_{\text{discrete}})$ is a Baire space.
 (iii) Consider $\mathbb{Q} \subseteq \mathbb{R}$ with its subspace topology. Then $\{q\} \subseteq \mathbb{Q}$ is a closed subset without inner points: an open subset of $\mathbb{Q}$ is of the form $\mathbb{Q} \cap \mathcal{O}$ with $\mathcal{O} \subseteq \mathbb{R}$ open. Hence an open subset of $\mathbb{Q}$ is either empty or has to contain infinitely many points. Now $\mathbb{Q} = \bigcup_{q \in \mathbb{Q}} \{q\}$ is a countable union of closed subsets without inner points. Hence by Proposition 7.1.5, (i), the rational numbers $\mathbb{Q}$ are *not* a Baire space.

We collect now some first properties of Baire spaces:

Proposition 7.1.8 *Let $(M, \mathcal{M})$ be a non-empty Baire space.*

 (i) *Let $\{A_n\}_{n \in \mathbb{N}}$ be a countable closed cover of M. Then at least one A_n has a non-empty open interior, $A_n^\circ \neq \emptyset$.*
 (ii) *Let $\mathcal{O} \subseteq M$ be a non-empty open subset then $(\mathcal{O}, \mathcal{M}|_{\mathcal{O}})$ is a Baire space again.*
 (iii) *Let $A \subseteq M$ be a meager subset. Then $(M \setminus A, \mathcal{M}|_{M \setminus A})$ is a Baire space again.*

Proof Let $\{A_n\}_{n \in \mathbb{N}}$ be a closed cover of M. Hence $M = \bigcup_{n \in \mathbb{N}} A_n$ contains inner points, in fact all points are inner, $M^\circ = M$. According to Proposition 7.1.5, (i), there must be at least one A_n with $A_n^\circ \neq \emptyset$. This shows the first part. For the second part let $\mathcal{O} \subseteq M$ be open and non-empty. Moreover, let $U_n \subseteq \mathcal{O}$ be open subsets which are dense in $\mathcal{O}$: we denote the closure inside $\mathcal{O}$ with respect to $\mathcal{M}|_{\mathcal{O}}$ with $\overline{U}_n$. Then $\overline{U}_n = \mathcal{O}$. Now consider $V_n = U_n \cup (M \setminus \mathcal{O}^{\text{cl}}) \subseteq M$ which is open since every open $U \subseteq \mathcal{O}$ is also open in M thanks to $\mathcal{O}$ being open. For the closure of V_n in M we get

$$
V_n^{\text{cl}} = \left(U_n \cup \left(M \setminus \mathcal{O}^{\text{cl}} \right) \right)^{\text{cl}} = U_n^{\text{cl}} \cup \left(M \setminus \mathcal{O}^{\text{cl}} \right)^{\text{cl}} = \mathcal{O}^{\text{cl}} \cup \left(M \setminus \mathcal{O}^{\text{cl}} \right)^{\text{cl}} = M,
$$

since $(M \setminus \mathcal{O}^{\text{cl}})^{\text{cl}}$ contains $M \setminus \mathcal{O}^{\text{cl}}$ and U_n being dense in $\mathcal{O}$ in the subspace topology implies that U_n is dense in $\mathcal{O}^{\text{cl}}$ in the original topology, see also Exercise 7.4.4. Since

M is a Baire space we know that $\bigcap_{n \in \mathbb{N}} V_n$ is still dense in M, by Proposition 7.1.5, (ii). Let us compute the closure of this intersection

$$\left(\bigcap_{n \in \mathbb{N}} V_n\right)^{\mathrm{cl}} = \left(\bigcap_{n \in \mathbb{N}} \left(U_n \cup \left(M \setminus \mathcal{O}^{\mathrm{cl}}\right)\right)\right)^{\mathrm{cl}}$$

$$= \left(\left(\bigcap_{n \in \mathbb{N}} U_n\right) \cup \left(M \setminus \mathcal{O}^{\mathrm{cl}}\right)\right)^{\mathrm{cl}}$$

$$= \left(\bigcap_{n \in \mathbb{N}} U_n\right)^{\mathrm{cl}} \cup \left(M \setminus \mathcal{O}^{\mathrm{cl}}\right)^{\mathrm{cl}}$$

$$= \left(\bigcap_{n \in \mathbb{N}} U_n\right)^{\mathrm{cl}} \cup \left(M \setminus \left(\mathcal{O}^{\mathrm{cl}}\right)^{\circ}\right), \qquad (*)$$

where we used Proposition 2.3.11, (v). Since $\mathcal{O}$ is open, it is clear that $\left(\mathcal{O}^{\mathrm{cl}}\right)^{\circ} \supseteq \mathcal{O}$. Thus the second contribution $M \setminus \left(\mathcal{O}^{\mathrm{cl}}\right)^{\circ}$ in $(*)$ can not contain any point from $\mathcal{O}$. Since the whole union is M since $\bigcap_{n \in \mathbb{N}} V_n$ is dense, we conclude

$$\mathcal{O} \subseteq \left(\bigcap_{n \in \mathbb{N}} U_n\right)^{\mathrm{cl}}. \qquad (**)$$

Let us use the fact that for every subset $A \subseteq \mathcal{O}$ in an open subset we have $\overline{A} = A^{\mathrm{cl}} \cap \mathcal{O}$, see Exercise 2.7.10. Back to $(**)$ we see that

$$\overline{\bigcap_{n \in \mathbb{N}} U_n} = \left(\bigcap_{n \in \mathbb{N}} U_n\right)^{\mathrm{cl}} \cap \mathcal{O} \overset{(**)}{=} \mathcal{O},$$

which finally shows that $\bigcap_{n \in \mathbb{N}} U_n$ is dense in $\mathcal{O}$ with respect to the subspace topology $\mathcal{M}|_{\mathcal{O}}$. Thus $\mathcal{O}$ is a Baire space by Proposition 7.1.5, (ii). For the third part, let $A \subseteq M$ be meager. Then we know from Proposition 7.1.5, (iv), that $M \setminus A$ is dense in M. Now suppose that $B \subseteq M \setminus A$ is meager in the subspace topology. Then $B = \bigcup_{n \in \mathbb{N}} B_n$ with $B_n \subseteq M \setminus A$ being nowhere dense in the subspace topology. Thus $B_n \subseteq M$ is nowhere dense as well, see Exercise 7.4.4, showing that B is meager in M, too. Hence the union $A \cup B$ is meager by Proposition 7.1.3, (iv), and thus $M \setminus (A \cup B)$ is dense in M by Proposition 7.1.5, (iv). But $M \setminus (A \cup B) = (M \setminus A) \cap (M \setminus B) = (M \setminus A) \setminus B$ is the complement of $B \subseteq M \setminus A$ inside $M \setminus A$. Being dense in M it is clearly dense in $M \setminus A$ as well. Thus Proposition 7.1.5, (iv), shows that $M \setminus A$ is a Baire space, too. $\qquad \square$

7.2 Baire's Theorem

Up to now we have not too many examples of Baire spaces as e.g. the constructions from Proposition 7.1.8 require to start with a Baire space. This changes now very much after showing the following two theorems:

Theorem 7.2.1 (Baire I) *A complete metric space* (M, d) *is a Baire space.*

Proof We want to apply the criterion from Proposition 7.1.5, (ii). Thus let $\{\mathcal{O}_n\}_{n \in \mathbb{N}}$ be a sequence of open dense subsets of M. If $U \subseteq M$ is an arbitrary non-empty open subset then $U \cap \mathcal{O}_n \neq \emptyset$ since $\mathcal{O}_n$ is dense, see Exercise 2.7.11. For $n = 1$ this shows that there is an open ball $\mathrm{B}_{\epsilon_1}(p_1) \subseteq \mathrm{B}_{\epsilon_1}(p_1)^{\mathrm{cl}} \subseteq U \cap \mathcal{O}_1$ for some small enough $\epsilon_1 > 0$ and some point $p_1 \in U \cap \mathcal{O}_1$. By induction, we can construct points $p_n \in U \cap \mathcal{O}_n$ and a zero sequence $\epsilon_n \longrightarrow 0$ such that $\mathrm{B}_{\epsilon_n}(p_n)^{\mathrm{cl}} \subseteq \mathrm{B}_{\epsilon_{n-1}}(p_{n-1}) \cap \mathcal{O}_n$ for all $n \geq 2$. This way we arrive at

$$\bigcap_{n=1}^{\infty} \mathrm{B}_{\epsilon_n}(p_n)^{\mathrm{cl}} \subseteq U \cap \bigcap_{n=1}^{\infty} \mathcal{O}_n. \tag{$*$}$$

From Exercise 7.4.5 we know that the completeness of (M, d) implies that the intersection on the left hand side in $(*)$ is non-empty. Hence the intersection on the right hand side is non-empty as well. Since U was an arbitrary open subset of M, the intersection $\bigcap_{n \in \mathbb{N}} \mathcal{O}_n$ is dense. Thus (M, d) is a Baire space.   □

This provides us a lot of interesting Baire spaces. The second version of Baire's Theorem adds yet more to the collection:

Theorem 7.2.2 (Baire II) *A locally compact Hausdorff space* $(M, \mathcal{M})$ *is a Baire space.*

Proof We proceed very similar to the metric case: let $\{\mathcal{O}_n\}_{n \in \mathbb{N}}$ be a sequence of open dense subsets of M. Let $U \subseteq M$ be a non-empty open subset. Then we know that $\mathcal{O}_1 \cap U$ is non-empty, too, since $\mathcal{O}_1$ is dense. Since we have a neighbourhood basis of compact neighbourhoods for every point by Corollary 5.4.3, we find an open subset $B_1 \subseteq M$ with

$$\emptyset \neq B_1 \subseteq B_1^{\mathrm{cl}} \subseteq \mathcal{O}_1 \cap U,$$

such that in addition B_1^{cl} is compact. Replacing U by B_1 and $\mathcal{O}_1$ by $\mathcal{O}_2$ etc. we get inductively a sequence of non-empty open subsets $\{B_n\}_{n \in \mathbb{N}}$ with compact closures B_n^{cl} such that

$$B_n^{\mathrm{cl}} \subseteq \mathcal{O}_n \cap B_{n-1}.$$

Now we consider the intersection

$$K = \bigcap_{n=1}^{\infty} B_n^{\mathrm{cl}},$$

which is closed (and even compact). We claim that K is non-empty. From $\cdots \subseteq B_n \subseteq B_n^{\mathrm{cl}} \subseteq B_{n-1} \subseteq \cdots \subseteq B_1^{\mathrm{cl}}$ we see that $U_n = B_1^{\mathrm{cl}} \setminus B_n^{\mathrm{cl}}$ is an open subset of B_1^{cl} in the subspace topology. Suppose that $K = \emptyset$ then this means that

$$\bigcup_{n=1}^{\infty} U_n = \bigcup_{n=1}^{\infty} \left(B_1^{\mathrm{cl}} \setminus B_n^{\mathrm{cl}} \right) = B_1^{\mathrm{cl}} \setminus \bigcap_{n=1}^{\infty} B_n^{\mathrm{cl}} = B_1^{\mathrm{cl}} \setminus K = B_1^{\mathrm{cl}}.$$

Since B_1^{cl} is compact, this open cover has a finite subcover. Since furthermore $U_n \subseteq U_{n+1}$ for all n we find some large $N \in \mathbb{N}$ with $U_N = B_1^{\mathrm{cl}}$. Thus $B_N^{\mathrm{cl}} = B_1^{\mathrm{cl}} \setminus U_N = \emptyset$ contradicts the construction of B_N^{cl}. We conclude that $K \neq \emptyset$. Now $K \subseteq B_n^{\mathrm{cl}} \subseteq \mathcal{O}_n \cap B_{n-1}$ shows that $K \subseteq \mathcal{O}_n$ for all n and thus K is also contained in the intersection of all $\mathcal{O}_n$. Hence $K \subseteq \bigcap_{n=1}^{\infty} \mathcal{O}_n$. Since also $K \subseteq U$ we see that

$$U \cap \bigcap_{n=1}^{\infty} \mathcal{O}_n \neq \emptyset.$$

Since U was an arbitrary open subset we see that $\bigcap_{n=1}^{\infty} \mathcal{O}_n$ has to be dense. Thus by the characterization of Baire spaces as in Proposition 7.1.5, (ii), the claim follows. $\square$

Clearly, we get a wealth of examples from these two theorems, some of which we list here:

Example 7.2.3 (Baire spaces II)

(i) The Euclidean $\mathbb{R}^n$ and open subsets of $\mathbb{R}^n$ are Baire spaces. Here we can argue with both versions of Baire's Theorem. Also $\mathbb{R} \setminus \mathbb{Q}$ is a Baire space by Proposition 7.1.8, (iii).

(ii) More generally, Banach spaces are Baire spaces. This has important consequences in functional analysis. Still more general, topological vector spaces which are metrizable are Baire spaces if they are complete. We have seen that $\mathscr{C}(M)$ is such an example for M being a second countable locally compact Hausdorff space.

(iii) All topological manifolds are Baire spaces by the second version of Baire's Theorem and Example 5.4.4, (iv).

As a first simple application we state the following proposition which can be seen as a quite general version of the "uniform boundedness principle":

Proposition 7.2.4 (Principle of uniform boundedness) *Let $(M, \mathcal{M})$ be a Baire space and let $\{f_i\}_{i \in I}$ be a set of continuous functions on M. Suppose that they are pointwise bounded on M. Then there exists a non-empty open subset $\mathcal{O} \subseteq M$ on which the f_i are uniformly bounded, i.e. there exists a $C > 0$ with*

$$\sup_{i \in I} \sup_{p \in \mathcal{O}} |f_i(p)| \leq C. \tag{7.2.1}$$

Proof For $n \in \mathbb{N}$ we consider the following subsets

$$A_n = \left\{ p \in M \mid |f_i(p)| \leq n \text{ for all } i \in I \right\}$$

of M, i.e. the intersection of all the pre-images $f_i^{-1}(B_n(0)^{cl})$ of the closed disk in $\mathbb{C}$ of radius n. Since the f_i are continuous, it follows that $A_n \subseteq M$ is closed. Then we see that $p \in A_n$ as soon as $\sup_{i \in I} |f_i(p)| \leq n$. Thus every point of M is in some A_n and we get

$$M = \bigcup_{n=1}^{\infty} A_n.$$

Since M is a Baire space, Proposition 7.1.8, (i), shows that at least one A_{n_0} has a non-empty open interior $\mathcal{O} = A_{n_0}^{\circ}$. Then taking this open subset and $C = n_0$ will do the job. $\square$

While this formulation may not seem very enlightening yet, there are serious applications of the Principle of Uniform Boundedness. Among them are the Banach-Steinhaus Theorem, the Open-Mapping Theorem, and the Closed-Graph Theorem. In some sense it is fair to say that these three theorems constitute the very basis of functional analysis.

7.3 Discontinuous Functions

We conclude this chapter with a short excursion to the world beyond continuity: it turns out that topological methods and in particular the techniques from Baire theory give some structure and nontrivial information also for discontinuous functions.

In many situations one is given a sequence $\{f_n\}_{n \in \mathbb{N}}$ of continuous functions on a topological space $(M, \mathcal{M})$ of which one knows pointwise convergence, i.e.

$$f(p) = \lim_{n \to \infty} f_n(p) \tag{7.3.1}$$

exists for all $p \in M$. This defines a new function on M. However, simplest examples from calculus show that f needs no longer to be continuous. Of course, adding some more conditions like locally uniform convergence will result in a continuous limit function, see e.g. Exercise 6.4.3. But what happens if one just has pointwise convergence alone? The quite surprising fact is that there are still reasonable statements one can make and the resulting function is far from being arbitrary. As a first step we have to characterize those points in M where the limit function is actually continuous:

Proposition 7.3.1 *Let* $(M, \mathcal{M})$ *be a topological space and let* $\{f_n\}_{n \in \mathbb{N}}$ *be a sequence of continuous functions such that*

$$f(p) = \lim_{n \to \infty} f_n(p) \tag{7.3.2}$$

exists for all $p \in M$. *For* $\epsilon > 0$ *let*

$$C_n(\epsilon) = \left\{ p \in M \mid |f(p) - f_n(p)| \le \epsilon \right\}, \tag{7.3.3}$$

and set

$$C(\epsilon) = \bigcup_{n=1}^{\infty} C_n^{\circ}(\epsilon) \quad and \quad C = \bigcap_{n=1}^{\infty} C\left(\tfrac{1}{n}\right). \tag{7.3.4}$$

Then f is continuous at $p \in M$ iff $p \in C$.

Proof Suppose first that f is continuous at $p \in M$ and let $\epsilon > 0$ be given. From the pointwise convergence (7.3.2) at p we get an $n \in \mathbb{N}$ with $|f(p) - f_n(p)| < \tfrac{\epsilon}{3}$. Since f as well as f_n are continuous at p we get an open neighbourhood $U \in \mathfrak{U}(p)$ with

$$|f(p) - f(q)| < \frac{\epsilon}{3} \quad and \quad |f_n(p) - f_n(q)| < \frac{\epsilon}{3}$$

for all points $q \in U$. For those points $q \in U$ this gives

$$|f(q) - f_n(q)| \le |f(q) - f(p)| + |f(p) - f_n(p)| + |f_n(p) - f_n(q)| < \epsilon.$$

This shows that $q \in C_n(\epsilon)$ and hence $U \subseteq C_n(\epsilon)$. But U is open and hence $p \in U \subseteq C_n^{\circ}(\epsilon)$ gives $p \in C(\epsilon)$. Since $\epsilon > 0$ was arbitrary, $p \in C$ follows. Conversely, let $p \in C$ and $\epsilon > 0$ be given. Then $p \in C(\tfrac{\epsilon}{3}) = U$. Taking this open neighbourhood of p gives by definition $|f(q) - f_n(q)| \le \tfrac{\epsilon}{3}$ for all $q \in C_n^{\circ}(\tfrac{\epsilon}{3})$. Since f_n is continuous, we have an open neighbourhood $V \in \mathfrak{U}(p)$ of p with $|f_n(q) - f_n(p)| < \tfrac{\epsilon}{3}$ for $q \in V$. For $q \in V \cap U$ this gives

$$|f(p) - f(q)| \le |f(p) - f_n(p)| + |f_n(p) - f_n(q)| + |f_n(q) - f(q)| < \epsilon.$$

showing the continuity of f at p since $V \cap U \in \mathfrak{U}(p)$ is still an open neighbourhood. $\qquad\square$

Proposition 7.3.2 *Let $(M, \mathcal{M})$ be a topological space and let $\{f_n\}_{n \in \mathbb{N}}$ be a sequence of continuous functions converging pointwise to a function f. Then the set of discontinuities of f is meager.*

Proof Using the notation of Proposition 7.3.1 we have to show that $M \setminus C$ is meager. Consider the closed subset

$$A_n(\epsilon) = \left\{ p \in M \mid |f_n(p) - f_{n+k}(p)| \le \epsilon \text{ for all } k \right\},$$

i.e. the intersection of the preimages of the closed disks $B_\epsilon(0)^{\mathrm{cl}}$ under the continuous functions $f_n - f_{n+k}$. Since $\lim_{n \to \infty} f_n(p) = f(p)$ for all $p \in M$ we have

$$\bigcup_{n=1}^{\infty} A_n(\epsilon) = M. \tag{$*$}$$

Moreover, for $p \in A_n(\epsilon)$ we have

$$|f_n(p) - f(p)| = \lim_{k \longrightarrow \infty} |f_n(p) - f_{n+k}(p)| \le \epsilon,$$

showing that $A_n(\epsilon) \subseteq C_n(\epsilon)$. Hence $A_n^\circ(\epsilon) \subseteq C_n^\circ(\epsilon)$, too, and thus

$$\bigcup_{n=1}^{\infty} A_n^\circ(\epsilon) \subseteq \bigcup_{n=1}^{\infty} C_n^\circ(\epsilon) = C(\epsilon). \qquad (**)$$

Since $A_n(\epsilon) = A_n(\epsilon)^{\mathrm{cl}}$ is closed its boundary $\partial A_n(\epsilon)$ belongs to $A_n(\epsilon)$. Thus the open interior of the boundary must be empty since $A_n(\epsilon) = \partial A_n(\epsilon) \cup A_n^\circ(\epsilon)$ and $A_n^\circ(\epsilon) \subseteq A_n(\epsilon)$ is the maximal open subset of $A_n(\epsilon)$. In particular $\partial A_n(\epsilon) = A_n(\epsilon) \setminus A_n^\circ(\epsilon)$ is nowhere dense. Hence the countable union $\bigcup_{n=1}^{\infty}(A_n(\epsilon) \setminus A_n^\circ(\epsilon))$ is a meager subset of M. From $(*)$ and $(**)$ we get

$$M \setminus C(\epsilon) \subseteq M \setminus \bigcup_{n=1}^{\infty} A_n^\circ(\epsilon) \subseteq \bigcup_{n=1}^{\infty}(A_n(\epsilon) \setminus A_n^\circ(\epsilon)).$$

Thus $M \setminus C(\epsilon)$ is the subset of a meager set and hence meager itself by Proposition 7.1.3, (iii). Since $C = \bigcap_{n=1}^{\infty} C\left(\frac{1}{n}\right)$ we see that

$$M \setminus C = \bigcup_{n=1}^{\infty} M \setminus C\left(\frac{1}{n}\right)$$

is a countable union of meager sets and hence again meager by Proposition 7.1.3, (iv). $\square$

For an arbitrary topological space M the above characterization does not tell much about the possible set of discontinuities since meager sets can be quite arbitrary. However, if the space M happens to be a Baire space then the statement is strong as meager sets in a Baire space are "small". We illustrate this with the following example:

Example 7.3.3 (Characteristic function of $\mathbb{Q}$) Consider the characteristic function $\chi_{\mathbb{Q}} \colon \mathbb{R} \longrightarrow \mathbb{R}$ of $\mathbb{Q} \subseteq \mathbb{R}$, i.e.

$$\chi_{\mathbb{Q}}(x) = \begin{cases} 1 & x \in \mathbb{Q} \\ 0 & x \in \mathbb{R} \setminus \mathbb{Q}, \end{cases} \qquad (7.3.5)$$

Since this function is well-known to be discontinuous at every point $x \in \mathbb{R}$ it can not be the pointwise limit of continuous functions.

Remark 7.3.4 (Baire classes). Consider $M = \mathbb{R}$ with its usual topology. Then the continuous functions on $\mathbb{R}$ form the zeroth Baire class H_0. The n-th Baire class

H_n are those functions which can be obtained as pointwise limits of functions in previous Baire classes but which are not already in $H_0, \ldots, H_{n-1}$. This defines the Baire classes H_n inductively. The example above show that $\chi_\mathbb{Q} \notin H_1$, in fact, one can show $\chi_\mathbb{Q} \in H_2$. It is a quite non-trivial theorem that all the Baire classes are non-empty and their union coincides with the space of measurable functions on $\mathbb{R}$ with respect to the usual Borel σ-algebra. More information on the Baire classes can be found e.g. in [10, Chap. IX, §4].

The next proposition characterizes the discontinuities of functions which are continuous on a dense subset:

Proposition 7.3.5 *Let* $(M, \mathcal{M})$ *be a topological space and let* $f : M \longrightarrow \mathbb{R}$ *be a function. If* f *is continuous on a dense subset of* M *then the set of discontinuities of* f *is meager.*

Proof Denote again the set of continuity points of f by $C \subseteq M$. For $n \in \mathbb{N}$ and $p \in C$ we have an open neighbourhood $U_n(p) \in \mathfrak{U}(p)$ such that

$$|f(p) - f(q)| < \frac{1}{n}$$

for $q \in U_n(p)$. Consider now $U_n = \bigcup_{p \in C} U_n(p)$ which is an open subset of M containing C. Since $C \subseteq U_n$ also U_n is dense. Thus $M \setminus U_n$ is closed and nowhere dense. Taking the union $\bigcup_{n=1}^\infty (M \setminus U_n)$ gives a meager subset. We want to show that $M \setminus C$ is contained in this subset. Thus let $p_0 \in M \setminus C$ be a point where f is discontinuous. This means that there is an $n \in \mathbb{N}$ such that for all open neighbourhoods $\mathcal{O}$ of p_0 there is a point $q_\mathcal{O} \in \mathcal{O}$ with $|f(p_0) - f(q_\mathcal{O})| \geq \frac{1}{n}$. This implies that $p_0 \notin U_{2n}$ since otherwise there would be a point $p \in C$ with $p_0 \in U_{2n}(p)$ and hence

$$|f(p_0) - f(q)| \leq |f(p_0) - f(p)| + |f(p) - f(q)| < \frac{1}{2n} + \frac{1}{2n} = \frac{1}{n}$$

for all $q \in U_{2n}(p)$. But this can not be possible for $q_\mathcal{O}$ with $\mathcal{O} = U_{2n}(p)$. Thus we conclude that the union $\bigcup_{n=1}^\infty (M \setminus U_n)$ contains $M \setminus C$ which therefore is meager as a subset of a meager set. $\square$

Corollary 7.3.6 *There is no function* $f : \mathbb{R} \longrightarrow \mathbb{R}$ *which is continuous on* $\mathbb{Q}$ *and discontinuous on* $R \setminus \mathbb{Q}$.

Proof The rationals $\mathbb{Q}$ are dense in $\mathbb{R}$ and hence any function f which is continuous on $\mathbb{Q}$ has a meager subset of discontinuities. Suppose f is discontinuous on $\mathbb{R} \setminus \mathbb{Q}$: Then $\mathbb{R} \setminus \mathbb{Q}$ would be meager in $\mathbb{R}$. Thus $\mathbb{R}$ is the countable (in fact finite) union of two meager subsets $\mathbb{R} = \mathbb{Q} \cup (\mathbb{R} \setminus \mathbb{Q})$ and hence meager itself. But this can not be the case for a Baire space e.g. by the characterization in Proposition 7.1.5, (iii). $\square$

Note that there are functions which are discontinuous on $\mathbb{Q}$ but continuous on $\mathbb{R} \setminus \mathbb{Q}$ like the function

$$f(x) = \begin{cases} 0 & x \in \mathbb{R} \setminus \mathbb{Q} \\ \frac{1}{m} & x = \frac{n}{m} \text{ for } n, m \text{ coprime.} \end{cases} \tag{7.3.6}$$

A last amusing fact on discontinuities is on differentiable functions and their derivatives:

Proposition 7.3.7 *Let $f : \mathbb{R} \longrightarrow \mathbb{R}$ be a differentiable function. Then the set of discontinuities of the derivative $f' : \mathbb{R} \longrightarrow \mathbb{R}$ is meager.*

Proof For all $x \in \mathbb{R}$ we obtain the value of the derivative $f'(x)$ as a pointwise limit

$$f'(x) = \lim_{n \longrightarrow \infty} f_n(x) \quad \text{where} \quad f_n(x) = n \left(f \left(x + \tfrac{1}{n} \right) - f(x) \right).$$

Since f_n is clearly continuous, the set of discontinuities of f' is meager by Proposition 7.3.2. $\square$

In particular, the set where f' is continuous is a dense subset of $\mathbb{R}$ by Proposition 7.1.5, (iv), and the fact that $\mathbb{R}$ is a Baire space. Moreover, in view of Remark 7.3.4, the derivative of a differentiable function is necessarily in the zeroth or first Baire class.

7.4 Exercises

Exercise 7.4.1 (A Baire space is not meager) Show that a Baire space is not meager in itself.

Exercise 7.4.2 (The Cantor set III) Show that, despite being a meager subset of $\mathbb{R}$, the Cantor set C is a Baire space itself.

Exercise 7.4.3 (Dense in an open subspace) Let $(M, \mathcal{M})$ be a topological space and let $\mathcal{O} \subseteq M$ be an open subset. Let furthermore $U \subseteq \mathcal{O}$ be dense in $\mathcal{O}$ with respect to the subspace topology. Show that U is dense in $\mathcal{O}^{\mathrm{cl}}$ with respect to the subspace topology of the closure $\mathcal{O}^{\mathrm{cl}}$ in M.

Exercise 7.4.4 (Nowhere dense in a subspace) Let $(M, \mathcal{M})$ be a topological space and let $N \subseteq M$ be a subset endowed with the subspace topology $\mathcal{M}\big|_N$ as usual. Suppose $A \subseteq N$ is nowhere dense in N with respect to $\mathcal{M}\big|_N$. Show that $A \subseteq M$ is nowhere dense in M, too.
Hint: Suppose $\mathcal{O} = (A^{\mathrm{cl}})^\circ$ is non-empty but $\mathcal{O} \cap N = \emptyset$. Why does this yield the desired contradiction?

Exercise 7.4.5 (Complete metric spaces) Let (M, d) be a metric space. Show that the following statements are equivalent:

(i) The metric space (M, d) is complete.
(ii) For every zero sequence $\epsilon_n \longrightarrow 0$ of positive numbers $\epsilon_n > 0$ and every sequence $p_n \in M$ such that $B_{\epsilon_{n+1}}(p_{n+1}) \subseteq B_{\epsilon_n}(p_n)$, the intersection $\bigcap_n B_{\epsilon_n}(p_n)$ is non-empty.

Hint: Suppose (M, d) is complete. Then show that the points $(p_n)_{n \in \mathbb{N}}$ form a Cauchy sequence in (M, d). The limit of this Cauchy sequence is the (unique) point in the intersection. The other direction is analogous.

Exercise 7.4.6 (The principle of uniform boundedness) Let $(M, \mathcal{M})$ be a Baire space and let $\{f_i\}_{i \in I}$ be a set of continuous functions on M. Suppose that they are pointwise bounded on M. Show that for every point $q \in M$ and every open neighbourhood $\mathcal{O} \subseteq M$ of this point q there exists a non-empty open subset $U \subseteq \mathcal{O}$ on which the f_i are uniformly bounded, i.e. there exists a $C > 0$ with

$$\sup_{i \in I} \sup_{p \in U} |f_i(p)| \leq C. \tag{7.4.1}$$

This is a refined version of Proposition 7.2.4. Note however, that U itself might not be a neighbourhood of q anymore.

Exercise 7.4.7 (The characteristic function of $\mathbb{Q}$) Prove that the characteristic function $\chi_{\mathbb{Q}}$ of the rational numbers $\mathbb{Q} \subseteq \mathbb{R}$ is in the second Baire class.
Hint: Fix an enumeration $\{q_n\}_{n \in \mathbb{N}}$ of the rationals and consider $\chi_n(x) = 1$ if $x = q_1, \ldots, q_n$ and $\chi_n(x) = 0$ else. Show that $\chi_n \longrightarrow \chi_{\mathbb{Q}}$ pointwise. How can you obtain each χ_n as the pointwise limit of continuous functions?

Exercise 7.4.8 (The space $\mathscr{C}_0(U)$) Let $U \subseteq \mathbb{R}^n$ be a non-empty open subset and consider the continuous functions $\mathscr{C}_0(U)$ with compact support inside U. For a compact subset $K \subseteq U$ we consider the continuous functions $\mathscr{C}_K(U)$ on U with support in K with the usual supremum norm, turning this into a Banach space. Consider all the seminorms p on $\mathscr{C}_0(U)$ for which the restriction $p|_{\mathscr{C}_K(U)}$ is a continuous seminorm on $\mathscr{C}_K(U)$, i.e. dominated by a multiple of the supremum norm over K. Consider now the topology on $\mathscr{C}_0(U)$ which is generated by all the open ϵ-balls $B_{p,\epsilon}(f)$ centered at functions $f \in \mathscr{C}_0(U)$ for all such seminorms p.

(i) Show that this (locally convex) topology is Hausdorff.
(ii) Show that the inclusion maps $\mathscr{C}_K(U) \longrightarrow \mathscr{C}_0(U)$ for all compact subsets $K \subseteq U$ are embeddings with closed images.
(iii) Show that $\mathscr{C}_0(U)$ is a countable union of $\mathscr{C}_{K_n}(U)$ for a suitable choice of compact subsets $K_n \subseteq U$.
(iv) Let $\mathcal{O} \subseteq \mathscr{C}_0(U)$ be an open neighbourhood of 0. Show that inside $\mathcal{O}$ one finds for every compact subset $K \subseteq U$ with $K^\circ \neq \emptyset$ a non-zero function $\varphi \in \mathscr{C}_K(U)$ with $\varphi \in \mathcal{O}$.
Hint: By definition, $\mathcal{O}$ contains an open ball $B_{p,\epsilon}(0)$ with a seminorm p having the above property. Show that for all K one can find a non-zero continuous function φ with $p(\varphi) < \epsilon$.

(v) Show that $\mathscr{C}_0(U)$ is not a Baire space. In particular, there is no complete metric on $\mathscr{C}_0(U)$ inducing the above topology.

This topology is of crucial importance in functional analysis. The same construction can also be done for functions with compact support which have better regularity properties than being just continuous: one is interested in $\mathscr{C}^k$ or even smooth functions. The resulting spaces $\mathscr{C}_0^k(U)$ with the analogous topology play then the role of test functions in the theory of distributions, see e.g. [14, 18, 19, 34] for a further reading.

Appendix A
Not an Introduction to Set Theory

In this short appendix we collect some basic information about set theory. Of course, this can not be a profound introduction to set theory at all. Instead, we assume that the reader has some familiarity with set theory as this can be acquired from basic courses in mathematics. Such a naive set theory will ultimately yield serious difficulties as this was the case also historically. The right answer to these problems is to pass to axiomatic set theory, which, however, would lead us much too far. Thus, this appendix should be more viewed as a *cheat sheet* in set theory rather than a systematic introduction. As a consequence, we only present some results without further proofs. Here the proofs are either simple and can be viewed as exercises, or they would require the full machinery of axiomatic set theory in order to be proven rigorously. An enjoyable further reading is e.g. [9].

A.1 Unions, Intersections, and Complements

In topology one needs to understand the behaviour of the set arithmetic operations, like unions and intersections, under taking images and preimages with respect to maps between the sets in question. We will collect now some basic features of set arithmetic.

In naive set theory, a set is a collection of "things", its *elements*, and two sets are equal if they have the same elements. If a "thing" x is an element of a set M we write this as $x \in M$, otherwise we write $x \notin M$. If M is a set, we can consider subsets of it: a certain collection of elements of M constitutes a *subset* N of M, written as $N \subseteq M$. Often we determine subsets by taking those elements of M satisfying a certain property p. In this case we write $N = \{x \in M \mid p(x) \text{ is true}\}$ for the subset of all those elements in M for which the statement $p(x)$ holds. In every set M there is a particular subset, the *empty set*, denoted by $\emptyset$, which has no elements at all. The collection of all subsets of a given set M forms again a set, called the *power set* of M, which we denote by 2^M.

© Springer International Publishing Switzerland 2014
S. Waldmann, *Topology*, DOI 10.1007/978-3-319-09680-3

It should be clear that up to now this can hardly be a serious mathematical definition as we have not said what a "collection" should be or what "things" are. However, from our current, naive point of view the ideas should become clear.

For a given non-empty set I we can consider other sets M_i for every $i \in I$. In this case we say that we have a collection $\{M_i\}_{i \in I}$ of sets indexed by an *index set* I. For any such collection of sets there is another set M, called the *union* of all the M_i, which contains the sets M_i as subsets but no other elements than those in the M_i. In this case we write $M = \bigcup_{i \in I} M_i$ or just $M = M_1 \cup \cdots \cup M_n$ if we have only finitely many sets $M_1, \ldots, M_n$. For the union of sets we have the following arithmetic rules

$$M \cup \emptyset = M = M \cup M, \tag{A.1.1}$$

and

$$M \cup N = N \cup M, \quad M \cup (N \cup K) = (M \cup N) \cup K, \tag{A.1.2}$$

for all sets M, N, and K.

Having a set M with a subset $N \subseteq M$ we can consider those elements of M which are not in N. This gives the *complement* of the subset N, written as

$$M \setminus N = \{x \in M \mid x \notin N\}. \tag{A.1.3}$$

If we have a collection $\{M_i\}_{i \in I}$ of sets indexed by some index set I we can form its *intersection*, i.e. all those elements which are in M_i for all $i \in I$. We denote this by

$$M = \bigcap_{i \in I} M_i = \{x \mid x \in M_i \text{ for all } i \in I\}. \tag{A.1.4}$$

If we have finitely many sets $M_1, \ldots, M_n$ we simply write $M = M_1 \cap \cdots \cap M_n$ for the intersection. It follows that $\bigcap_{i \in I} M_i \subseteq M_j$ for every $j \in I$.

We collect now some simple arithmetic rules for the complement and the intersection. For a set M with a subset $N \subseteq M$ we have

$$M \setminus M = \emptyset, \quad M \setminus \emptyset = M, \quad \text{and} \quad M \setminus (M \setminus N) = N. \tag{A.1.5}$$

Analogously to the union we get for the intersection

$$M \cap \emptyset = \emptyset, \quad M \cap M = M, \tag{A.1.6}$$

and

$$M \cap N = N \cap M, \quad M \cap (N \cap K) = (M \cap N) \cap K, \tag{A.1.7}$$

for all sets M, N, and K.

For the relations between unions, intersections, and complements we note the following distributive like rules. For arbitrary collections of sets $\{M_i\}_{i \in I}$ and $\{N_j\}_{j \in J}$ we have

$$\left(\bigcap_{i \in I} M_i\right) \cup \left(\bigcap_{j \in J} N_j\right) = \bigcap_{i \in I, j \in J} (M_i \cup N_j), \qquad (A.1.8)$$

and

$$\left(\bigcup_{i \in I} M_i\right) \cap \left(\bigcup_{j \in J} N_j\right) = \bigcup_{i \in I, j \in J} (M_i \cap N_j). \qquad (A.1.9)$$

For a collection $\{M_i\}_{i \in I}$ of subsets of a common set M we get

$$M \setminus \left(\bigcup_{i \in I} M_i\right) = \bigcap_{i \in I} (M \setminus M_i) \qquad (A.1.10)$$

and

$$M \setminus \left(\bigcap_{i \in I} M_i\right) = \bigcup_{i \in I} (M \setminus M_i). \qquad (A.1.11)$$

The last important construction is the *Cartesian product* M of a collection of sets $\{M_i\}_{i \in I}$. The construction of finite Cartesian products requires ordered pairs and then ordered n-tupels, for infinite Cartesian products the construction is slightly more involved. The idea is however rather simple, an element x of the Cartesian product M is a large column $(x_i)_{i \in I}$ indexed by I such that at the i-th position we have an element $x_i \in M_i$. We write $M = \prod_{i \in I} M_i$ for the Cartesian product and $M = M_1 \times \cdots \times M_n$ for finitely many $M_1, \ldots, M_n$.

A.2 Maps, Images, and Preimages

Given two sets M and N a *map* from M to N is an assignment where every element $x \in M$ is assigned an element $f(x) \in N$. In this case we write $f : M \ni x \mapsto f(x) \in N$ for the map f and call $f(x)$ the image of x under the map f. A particular map is the *identity map* on M which maps every element $x \in M$ to x itself. We denote the identity map on M by $\mathrm{id}_M : M \longrightarrow M$.

For a subset $A \subseteq M$ one defines the image set

$$f(A) = \{f(x) \in N \mid x \in A\}, \qquad (A.2.1)$$

and for a subset $B \subseteq N$ one defines the preimage of B by

$$f^{-1}(B) = \{x \in M \mid f(x) \in B\}. \qquad (A.2.2)$$

This way, a map $f : M \longrightarrow N$ induces a map $f : 2^M \longrightarrow 2^N$, the image map, and a map $f^{-1} : 2^N \longrightarrow 2^M$, the preimage map. We clearly have $f^{-1}(N) = M$ and

$f^{-1}(\emptyset) = \emptyset$, as well as $f(\emptyset) = \emptyset$. However, $f(M)$ needs not to be N. For unions and intersections we have the following properties

$$f\left(\bigcap_{i \in I} A_i\right) \subseteq \bigcap_{i \in I} f(A_i) \tag{A.2.3}$$

and

$$f\left(\bigcup_{i \in I} A_i\right) = \bigcup_{i \in I} f(A_i) \tag{A.2.4}$$

for every collection $\{A_i\}_{i \in I}$ of subsets $A_i \subseteq M$. Note that the inclusion in (A.2.3) can be proper. Note also that there is no simple behaviour of complements with respect to images. For the preimage we get

$$f^{-1}(N \setminus B) = M \setminus f^{-1}(B), \tag{A.2.5}$$

$$f^{-1}\left(\bigcap_{i \in I} B_i\right) = \bigcap_{i \in I} f^{-1}(B_i), \tag{A.2.6}$$

and

$$f^{-1}\left(\bigcup_{i \in I} B_i\right) = \bigcup_{i \in I} f^{-1}(B_i) \tag{A.2.7}$$

for a subset $B \subseteq N$ and for all collections $\{B_i\}_{i \in I}$ of subsets $B_i \subseteq N$.

Having two maps $f: M \longrightarrow N$ and $g: N \longrightarrow K$ one defines their composition $g \circ f: M \longrightarrow K$ by

$$(g \circ f)(x) = g(f(x)) \tag{A.2.8}$$

for $x \in M$. The composition is associative and we have $\mathrm{id}_N \circ f = f = f \circ \mathrm{id}_M$.

The composition behaves well with preimages: we have the crucial rule

$$\mathrm{id}_M^{-1} = \mathrm{id}_{2^M} \quad \text{and} \quad (g \circ f)^{-1} = g^{-1} \circ f^{-1}. \tag{A.2.9}$$

A map $f: M \longrightarrow N$ is called *injective*, if every element $y \in N$ has at most one preimage in M. The map is called *surjective* if every element $y \in N$ is an image, i.e. has a preimage in M. Equivalently, this means $f(M) = N$. Finally, f is called *bijective* if f is both injective and surjective. Being bijective is equivalent to having an *inverse map*, i.e. a map denoted by $f^{-1}: N \longrightarrow M$ with $f \circ f^{-1} = \mathrm{id}_N$ and $f^{-1} \circ f = \mathrm{id}_M$. Unfortunately, we have here a certain clash of notations from the same symbol for the inverse map and the preimage map.

A.3 Zorn's Lemma

In this last section we present a formulation of the Axiom of Choice in set theory in the incarnation of Zorn's Lemma. Since we have not discussed the other axioms of set theory properly, there is no point in proving that Zorn's Lemma is equivalent to the Axiom of Choice under the assumption of the remaining axioms. We refer to [11] for a detailed discussion.

The Axiom of Choice states that for an arbitrary collection of non-empty sets $\{M_i\}_{i \in I}$ also their Cartesian product $\prod_{i \in I} M_i$ is non-empty. The name originates from the idea that the elements of the Cartesian product $x = (x_i)_{i \in I}$ can be viewed as *choice functions*, i.e. for every index $i \in I$ the element x "chooses" the element $x_i \in M_i$. Though seemingly obvious, the subtle point with this axiom of set theory is that we have to be able to choose elements for all i from a priori completely structureless sets M_i. As soon as the sets M_i are more specific, this might be possible without invoking the Axiom of Choice.

The formulation of the Axiom of Choice we need most in this work is Zorn's Lemma. Here one considers a partially ordered set M with partial order $\preccurlyeq$. A subset $N \subseteq M$ is called *linearly* (or *totally*) ordered with respect to $\preccurlyeq$ if any two elements $x, y \in N$ are comparable with respect to $\preccurlyeq$, i.e. either $x \preccurlyeq y$ or $y \preccurlyeq x$. A linearly ordered subset is also called a *chain*. A subset N has an *upper bound* $x \in M$ if for all $y \in N$ we have $y \preccurlyeq x$. An element $x_\infty \in M$ is called *maximal* if for all $x \in M$ with $x_\infty \preccurlyeq x$ one has $x_\infty = x$. With these notions, Zorn's Lemma can be formulated as follows:

Lemma A.3.1 (Zorn) *Let $(M, \preccurlyeq)$ be a partially ordered set such that every linearly ordered subset N has an upper bound in M. Then M has a maximal element with respect to $\preccurlyeq$.*

The validity of Zorn's Lemma for all partially ordered sets is then equivalent to the Axiom of Choice.

Note that the maximal element can in general not be obtained in any algorithmic way. It will also not be unique in general. Zorn's Lemma gives only the existence, nothing more and nothing less.

References

1. Arveson, W.: An invitation to C^*-algebras. Graduate Texts in Mathematics, vol. 39. Springer, New York (1998)
2. Barner, M., Flohr, F.: Analysis I, 3rd edn. Walter de Gruyter, New York (1987)
3. Bourbaki, N.: General topology. Chapters 1–4. In: Elements of Mathematics. Springer, Berlin (1998). Reprint of the 1989 English translation
4. Bredon, G.E.: Topology and geometry. Graduate Texts in Mathematics, vol. 139. Springer, New York (1993)
5. Bredon, G.E.: Sheaf theory, 2nd edn. Graduate Texts in Mathematics, vol. 170. Springer, New York (1997)
6. Bröcker, T., Jänich, K.: Einführung in die Differentialtopologie, Heidelberger Taschenbücher, vol. 143. Springer, New York (1990). Korrigierter Nachdruck
7. Dixmier, J.: C^*-Algebras. North-Holland Publishing Co., Amsterdam (1977). Translated from the French by Francis Jellett, North-Holland Mathematical Library, vol. 15
8. Fulton, W.: Algebraic topology. Graduate Texts in Mathematics, vol. 153. Springer, New York (1995)
9. Halmos, P.R.: Naive Set Theory. Springer, Heidelberg, New York (1974). Reprint of the 1960 edition, Undergraduate Texts in Mathematics
10. Hausdorff, F.: Grundzüge der Mengenlehre. Chelsea Publishing Company, New York (1949)
11. Herrlich, H.: Axiom of choice. In: Lecture Notes in Mathematics, vol. 1876. Springer, Berlin (2006)
12. Hirsch, M.W.: Differential topology. Graduate Texts in Mathematics, vol. 33. Springer, New York (1976)
13. Jänich, K.: Topologie, 4th edn. Springer, New York (1994)
14. Jarchow, H.: Locally Convex Spaces. B. G. Teubner, Stuttdart (1981)
15. Kadison, R.V., Ringrose, J.R.: Fundamentals of the theory of operator algebras. Volume I: elementary theory. Graduate Studies in Mathematics, vol. 15. American Mathematical Society, Providence (1997)
16. Kadison, R. V., Ringrose, J. R.: Fundamentals of the theory of operator algebras. Volume II: advanced theory. Graduate Studies in Mathematics, vol. 16. American Mathematical Society, Providence (1997)
17. Kelley, J.L.: General topology. Graduate Texts in Mathematics, vol. 27. Springer, New York (1955)
18. Köthe, G.: Topological Vector Spaces I. Grundlehren der mathematischen Wissenschaft no. 159. Springer, New York (1969)

© Springer International Publishing Switzerland 2014
S. Waldmann, *Topology*, DOI 10.1007/978-3-319-09680-3

19. Köthe, G.: Topological Vector Spaces II. Grundlehren der mathematischen Wissenschaft no. 237. Springer, New York (1979)
20. Lang, S.: Real and functional analysis, 3rd edn. Graduate Texts in Mathematics, vol. 142. Springer, New York (1993)
21. Lang, S.: Fundamentals of differential geometry. Graduate Texts in Mathematics, vol. 191. Springer, New York (1999)
22. Lee, J.M.: Introduction to topological manifolds. Graduate Texts in Mathematics, vol. 202. Springer, New York (2000)
23. Lee, J.M.: Introduction to smooth manifolds. Graduate Texts in Mathematics, vol. 218. Springer, New York (2003)
24. MacLane, S.: Categories for the working mathematician, 2nd edn. Graduate Texts in Mathematics, vol. 5. Springer, New York (1998)
25. Michor, P.W.: Topics in differential geometry. Graduate Studies in Mathematics, vol. 93. American Mathematical Society, Providence, RI (2008)
26. Palais, R.S.: When proper maps are closed. Proc. Am. Math. Soc. **24**, 835–836 (1970)
27. von Querenburg, B.: Mengentheoretische Topologie, 3rd edn. Springer, New York (2001)
28. Rudin, W.: Real and Complex Analysis, 3rd edn. McGraw-Hill Book Company, New York (1987)
29. Rudin, W.: Functional Analysis, 2nd edn. McGraw-Hill Book Company, New York (1991)
30. Sakai, S.: C^*-Algebras and W^*-algebras. In: Classics in Mathematics. Springer, New York (1998). Reprint of the 1971 edition
31. Spanier, E.H.: Algebraic Topology. Springer, New York (1989). First corrected Springer edition, Softcover
32. Steen, L.A., Seebach, Jr., J.A.: Counterexamples in Topology. Dover Publications Inc., Mineola, NY (1995). Reprint of the 2nd (1978) edition
33. tom Dieck, T.: Algebraic topology, 2nd edn. In: EMS Textbooks in Mathematics. European Mathematical Society (EMS), Zürich (2010)
34. Treves, F.: Topological Vector Spaces. Distributions and Kernels. Academic Press, New York, London (1967)
35. Werner, D.: Funktionalanalysis, 4th edn. Springer, New York (2002)

Index

Printed by Printforce, the Netherlands